Running Wire at the Front Lines

Running Wire at the Front Lines

Memoir of a Radio and Telephone Man in World War II

Louis J. Lauria

Edited by Amanda Page Anderson

McFarland & Company, Inc., Publishers
Jefferson, North Carolina, and London

LIBRARY OF CONGRESS CATALOGUING-IN-PUBLICATION DATA

Lauria, Louis J.
    Running wire at the front lines : memoir of a radio and telephone man in World War II / Louis J. Lauria ; edited by Amanda Page Anderson.
        p.   cm.
    Includes bibliographical references and index.

    ISBN 978-0-7864-5926-1
    softcover : 50# alkaline paper ∞

    1. Lauria, Louis J. 2. World War, 1939–1945—Campaigns—Western Front. 3. World War, 1939–1945—Personal narratives, American. 4. United States. Army. Infantry Regiment, 11th. 5. United States. Army—Radiomen—Biography. 6. Radio operators—United States—Biography. 7. Soldiers—United States—Biography. I. Anderson, Amanda Page, 1979– II. Title.
    D756.L38   2011
    940.54'1273092—dc22   2010045065

British Library cataloguing data are available

© 2011 Olga Lauria and Amanda Page Anderson. All rights reserved

No part of this book may be reproduced or transmitted in any form or by any means, electronic or mechanical, including photocopying or recording, or by any information storage and retrieval system, without permission in writing from the publisher.

On the cover: Louis J. Lauria during World War II (courtesy of the author's estate); military field equipment © 2011 photos.com

Manufactured in the United States of America

McFarland & Company, Inc., Publishers
  Box 611, Jefferson, North Carolina 28640
    www.mcfarlandpub.com

Dedicated to the true heroes
of the Fifth Infantry Division
who made the supreme sacrifice
and gave their lives for others' freedom

Table of Contents

Preface	1
Introduction	7
I. Going into the Army	19
II. Overseas	26
III. Normandy	31
IV. Angers	49
V. Chartres	62
VI. Advancing through France	70
VII. Dornot	77
VIII. Metz	89
IX. Battle of the Bulge	101
X. Sauer River Crossings	111
XI. Bitburg	125
XII. Lieg	135
XIII. Oppenheim: Crossing the Rhine	143
XIV. Main River Bridge in Frankfurt	149
XV. End of the War	158
XVI. Home	165
Afterword	173
Appendix: The Fifth Infantry Division in World War II	179
Chapter Notes	207
Works Cited	213
Index	215

Preface
Amanda Page Anderson

What I remember most about my grandfather were his hands. He was a small man but he had the hands of a giant. When we danced together at his 60th wedding anniversary party, his hand engulfed mine and two of his fingers covered my wrist. His hands were worn and tough from a lifetime of hard labor.

I can still picture tracing my fingers along the massive veins crisscrossing the top of his hands when I was a little girl. His strong hands lifted me up on his knee to sit with him or snuck me a piece of candy when my parents weren't looking. He was a talented cartoonist and his giant hands swallowed a pen to draw cartoons of Popeye and Olive or Donald Duck for my sister and me.

My grandfather first started to record his experiences in World War II through his illustrations. A few years after the war, my grandfather drew from memory two large 11" x 17" pictures of German SS soldiers. He framed them and they hung in his furnished basement for almost 50 years. I wonder why he chose to draw the Germans first. Perhaps their images were the hardest to erase. He was a spotter for his Cannon Company and maybe he felt responsible for those German soldiers' deaths.

Over the years, he began to draw more and more images of the war. I remember him calling me over to show me some of them. I'll never forget when I first saw the sketch of his friends who were blown apart by shells. His illustrations depicted particular events he experienced during combat.

In the early 1990s, my grandfather began to write the stories to go along with his illustrations. It was then he started to talk more openly about the war. He would wait to make sure you were interested before he began to share. He was a quiet man but loved to talk once you and he were alone and he knew he had your attention. During holidays, when

One of the first illustrations that my grandfather drew of the war. It is of German SS troopers.

his house was filled with commotion from an endless stream of relatives, I would seek him out in a quiet corner and eagerly listen to tales about his experience growing up with 9 brothers and sisters or about the war. His memory was truly remarkable and family members would always defer to him when they had questions about the past.

He wrote about the war in spiral notebooks and sometimes on small notepads. Despite only having a sixth-grade education, he was compelled to record his memories. He sat at his kitchen table with a dictionary and a cup of coffee, writing. He shared his stories with his wife Olga, who he loved and treated with nothing but respect and kindness.

Writing was a therapeutic way for him to deal with the strong emotions he had carried with him his whole life. He witnessed the violent deaths of many of his friends in combat and at the end of the war suffered a nervous breakdown. He never received the medical care he needed. He checked himself out of the hospital in France early so he could meet up with his company, which was leaving for the States. He desperately wanted to return home.

Ever since I was little, I loved to listen to his stories. For my undergraduate thesis at Davidson College, I researched the children of Italian immigrants and interviewed my grandfather on multiple occasions. I was able to use these interviews to help write an introduction to his memoir about his childhood. I was fascinated by his past. In the late 1990s, my grandfather asked me if I would type up some of his war journals. Over the years, I typed up many different notebooks that he gave me. He would ask me what stories he had already given me so he could see if he left out any parts. He reviewed some of what I typed and made some corrections to his work. He learned how to use a computer and began typing up a few stories on his own. A regular keyboard was too small for him and he had to use a keyboard with oversized keys. His enormous fingers hit several keys at once on a standard keyboard.

My grandfather passed away on November 8, 2008. His death has left those of us who knew and loved him with a profound sense of empti-

My grandfather and I dancing at his 60th wedding anniversary party. His enormous hands completely engulfed mine.

ness. After he died, I began to compile everything I had typed over the past ten years. I wanted to make sure nothing was lost. All in all there were about nine different notebooks filled with stories. I asked my grandmother to ship me all of his journals and then began to piece together one cohesive memoir.

At first, I set out to make copies for my family so we would all be able to cherish his work. I also planned on sending a copy to a World War II archive. After everything had been put in order, the story flowed from his induction to the end of the war. It was compelling. His honest description of his combat experiences brings them to life. Family members who read his memoir couldn't put it down. There are very few World War II memoirs written by Italian Americans, especially by those from poor working-class families. I know my grandfather would have been overjoyed to know his memoir was published.

This is a German SS trooper my grandfather remembered from crossing over the Sauer River into Germany.

When compiling and editing this memoir, I made sure to preserve my grandfather's voice throughout the book, so I kept his words and only corrected blatant errors. Once, when we went over some of his typed stories together, he asked me if they were "pretty good for a sixth grade education." I said they were. I'm amazed at what he was able to accomplish. The margins of his notebooks are filled with his notes. He wrote most of the stories down multiple times in his notebooks. If he left something out, he would rewrite the whole story. The

Louis Lauria (center) in 2001 at the award ceremony where he was awarded the Jubilee Medal of Honor.

duplicate stories were very helpful because to each version was added a little bit more information. The facts and events always remained the same and I can say with strong confidence that he didn't embellish. Later, I thoroughly researched the actions of the Fifth Infantry Division to include in my grandfather's memoir and his recollections accurately coincide with the division's history.

I compiled his stories to include all of his words and focused on preserving his thoughts rather than streamlining the work. He had an excellent memory for dates and locations, which made organizing the book a lot easier. I researched the actions of the Fifth Infantry Division in the European Theater of Operations (ETO) and pieced his stories together in chronological order.

My only regret is not completing this memoir while he was still alive. Many times while working on it, I wished I could ask him questions. I found his notes on scraps of paper in between the pages of his division book from 1946 explaining that he was standing to the left of where a picture was taken or that he saw the bodies of dead GI in a particular photograph. He also left notes between the pages of other books and wrote on the back of his war photographs. I like to think he left those notes for me so I could finish his memoir. He was very thorough and provided a wealth of information about his life in combat. The names are all accurate and I was able to reference them with draft records, lists of commanding officers in the Fifth Infantry Division, and the roster for the Eleventh Infantry Regiment.

    I fondly remember my grandfather giving me a wink as he told me to make his book more interesting. I didn't add anything to his stories. I didn't need to. I'm grateful that he took the time to write down his memories. He left us with something wonderful to remember him by and I will always cherish his work. I'm thrilled to share his memoir with the public. My grandfather would have been proud. I am truly blessed that I can hear his voice again whenever I pick up his memoir.

Introduction

Louis J. Lauria was born May 19, 1924, to Angelina and Anthony Lauria. He grew up in Brooklyn, New York, and was one of ten children. His parents emigrated from Marsico Nuovo, a small village in the province of Potenza of the southern Italian region of Basilicata. Nestled in the mountains, Marsico Nuovo is a small, isolated, rural town. During the turn of the 20th century, southern Italy was plagued with poverty, hunger, disease and recurrent natural disasters, causing thousands of Italians to immigrate to America and seek a better life. Louis's mother, Angelina Azzato, was born in the United States, but her parents returned with her to Marsico Nuovo when she was young. While in Italy, she met Anthony Lauria and they married in 1908. Their first child, a girl, died as toddler. They had their first son, Ralph, in 1911. Angelina hoped to return to the United States and join family members who had settled in Brooklyn.

Anthony came to America for a job around 1916 and left Angelina and Ralph in Italy while he worked for the New York City department of sanitation. He lived with his sister Rose in Brooklyn and saved money in order to bring Angelina and Ralph over to the United States and buy a home. When World War I broke out, he fought for the U.S. After the war, he returned to the U.S. and sent for his wife and son to join him. Anthony praised America and was happy to have a steady job and the ability to feed his family. Life in southern Italy was full of hardships and in the United States at least the family could find work.

Like thousands of other Italian immigrants, Anthony was an unskilled laborer.[1] Since he grew up with poverty all around him, he felt fortunate to find a steady job with the sanitation department. Initially, Anthony collected garbage with his own horse and wagon. Later on, trucks were used to collect the garbage. The garbage pails were very heavy and often filled with coal from the coal-burning stoves that families used to heat their homes at that time.[2]

After several years of lifting the heavy cans, Anthony developed a double hernia from the hard labor and was transferred to the department of street cleaners. He worked as a street cleaner with a broom, a shovel and a can on wheels, keeping the streets of Brooklyn clean until he was 56 years old. He made friends with the butchers who had shops on the streets he cleaned and during his lunch breaks he would sweep up their shops. The butchers gave him gifts of meat which helped him feed his family of 12 during the Great Depression. They continued to bring him gifts long into his retirement.

Nine years after the birth of their first son Ralph, Anthony and Angelina had their daughter Kate in 1920 and their second oldest son Matthew in 1922. Louis was born on May 19, 1924, followed by his brothers August in 1926, Carmine in 1927, John in 1929, Anthony in 1930 and Daniel in 1932. The youngest girl, Rosemary, was born in 1934, making 10 living Lauria children in all.

The Lauria family settled in Brooklyn, NY, along with other Italian immigrants from Marsico Nuovo. They tended to stick together with people from their hometown and didn't associate much with Italians from other villages.[3] They moved into a four-family house at 101 Walworth Street. The Lauria family lived on the fourth floor and their Aunt Rose lived on the second floor. They rented the first and third floors out to other families.

The family worked hard to make ends meet. Anthony brought home five dollars a week from his job with the department of sanitation. From a young age, the children found odd jobs to chip in and help the family. When Louis was just five years old, he would go down to the market to get vegetables and help the farmers who drove their trucks in from the country. The farmers spent all day there selling their goods and he used to get work helping them. He would make 10 or 15 cents a day helping them sell their produce.[4] When he was a little older, he and his brothers would go to Wallabout market near the Brooklyn navy yard. They would help the farmers fix crates and clean up around their trucks. They earned around 25 cents for a few hours of work. Louis usually used his money to help the family but sometimes spent some of his earnings on the movies or candy.

Louis kept busy from a young age helping the family bring in extra money. He sold bottles for the two-cent deposit fee and searched for scrap iron in junkyards so he could sell it for one cent a pound. He also bought bushels of pretzels for 25 cents and sold them on the street for a profit.

On Saturdays and Sundays, the boys would shine shoes. When Louis was old enough, he got his own shoe-shine box and on a good day could make 15 or 20 cents.

The Lauria children all had jobs to do in the house. The boys were in charge of maintaining the family's coal stove. They had to sift ashes, bring wood up from the cellar, get rid of the ashes, and clean the coal stove with black liquid shoe polish to make the stove look nicer. Louis ironed the family's shirts and was teased by his brothers because he over-starched their collars. Kate and Rosie helped their mother with the endless cooking and cleaning.

Louis was known by his siblings as willing to help everybody. He would bring his father lunch every day and walk to Wallabout market. He was the only boy who helped his sisters and mother with the housework.

The family had a vegetable garden and used to make everything by hand. Louis explained, "God forbid they buy canned food out of the store. Everything had to be made by them. My father used to grow it. My mother would jar it. When you went down into our basement, there would be jars and jars of vinegar peppers and sausages. Whatever they wanted they would make it themselves."[5] Louis recalled that he loved when his mother made scrambled eggs with dried peppers from their garden and eggs from their own chickens.[6] His father made his own wine, which was renowned for its strength rather than its taste. The family helped make the wine from 20-pound boxes of grapes they bought at the market. They had 50-gallon barrels of wine in the basement where the boys would help their father crush the grapes and ferment the wine.[7]

One day, when Louis was around eight years old, he was helping his mother make sausages when he cut off the tip of his right middle finger. From then on, the first three fingers on his right hand were even with each other.

The Lauria family was Roman Catholic and all the children went to church without question. They had to learn the catechism by heart. The nuns used to have the children repeat every prayer that was in the book. If they failed, they had to go through the whole process again for another year.[8] Louis's younger sister, Rose, explained, "We were Catholic, that's what we were. You did what you had to do and that was it. There was no trying something else or not doing anything. It was a staple. That is what you did and that was it. You got a certain age—eight years old you made your communion, eleven you made your confirmation."[9]

Like other Italian immigrants from southern Italy, they brought with them customs and superstitions of their own culture. It wasn't uncommon to use garlic to ward off sickness. Louis's mother used many home remedies she grew up with in southern Italy. When her children hurt themselves, she made a mustard plaster which was a poultice of mustard seed powder wrapped in a cloth. She would wrap the injury in the cloth to help it heal. When her children had sore throats, she warmed up salt in a nylon stocking and wrapped it around their necks.[10] For ringworm, she used kerosene and vinegar, and if the children got warts she would tie a string around them and soak them in iodine.[11]

The family also believed in Mal'occhio, the "evil eye," which refers to the power of a person to invoke a curse with a dirty look. The curse usually manifested in a bad headache or recurrent bad luck. If they thought someone was under the curse of the "evil eye," they would hold a bowl of water over the person's head and drop three drops of oil into the water. If the oil coagulated, it meant the person had the "evil eye." If the sufferer was cursed, they recited a chant to ward off the evil. One time, Anthony performed the ritual for John when he had a bad headache. John's curse was so strong that it made his father throw up when he tried to do the ritual. After that, his father refused to do the ritual again.[12]

Louis and his brothers found inventive ways to make money and get ahead. In those days, each street had a regular hot-dog vendor. For five cents, you could get a frankfurter and a little cup of lemonade. Walworth Street had an elderly

This is Louis with his godfather, Joseph Logerotta, at his holy communion. Louis' father and Joseph's father were close friends in Italy.

hot-dog vendor, Nick, who liked his wine. He spent a good deal of time in the local grocery shop with a glass of wine, visiting with the "old men's club."[13] Afterward, Nick liked to take a nap by his cart.

One day, Louis and his friend Al told Nick that he could sleep while they watched his wagon. So, Nick fell asleep on the stoop. A bunch of Louis's brothers and friends came by and Al offered them a hot dog. He didn't charge his buddies a cent. This went on for about an hour while Nick slept. A few more kids came by and asked Al if he had any hot dogs left. Al said, "Yeah, I think I got one left on the bottom." He got the fork and tried a few times to grab the last hot dog. He finally scooped it up and saw that it was Nick's false teeth. Nobody wanted to eat hot dogs anymore after that.[14] Some people on the street didn't believe the boys, but those who were there that day never went to Nick's hot dog cart again.[15]

The Lauria boys knew the ins and outs of their street and never missed an opportunity to get ahead. Le Cody Pie Company had about 20 trucks that delivered pies to restaurants. At the end of the day, six or seven trucks would park in a garage that was on Walworth Street. Some trucks would bring back pies that hadn't sold, so the boys would sneak into the garage and check out all the trucks. They would usually find pies of all sorts of varieties: apple, pineapple, cheese, and blueberry. Most days their efforts paid off and they were rewarded with at least one pie.

When Louis was 11, his father got him a small white dog. She was a nice, obedient dog and smart. Louis raised her, named her Beauty, and trained her to do tricks. Louis and his brothers loved Beauty. When Beauty was about four years old, she got very sick. She didn't want to eat for a couple of days and eventually died. Louis cried for days and was devastated. It turned out that his Aunt Rose hadn't liked the dog. She had asked Louis's father multiple times to get rid of her but he refused. Rose wanted to get rid of the dog so badly that she poisoned it. Louis was heartbroken but managed to forgive his aunt and move on.

Louis eventually got another dog that was quite different from his sweet little Beauty. He wrote the following story about the experience:

> At the age of 13, I lived on Walworth Street in Brooklyn. My father owned an apartment with his sister and we lived on the fourth floor and she lived on the second floor. My father had about fifteen chickens in the backyard. They lived in a wood shanty where we stored our firewood for the winter months.
> In the back of the yard, there was a house that belonged to the next street

which was rented to a black lady by the name of Miss Keeps. My father always gave her some of the food from our garden, eggs from our chickens, and also some wine that he made every year. She was glad to get it. My father had been giving her food for many years. There were times my father would tell me to go down to Mrs. Keeps and bring her food. I was too short to reach her window, which was about 7 feet high. I used to get a long stick and tap on the window until she came over. I would get on a box and hand her the food and she would bend down to get it.

Every day someone in the family would feed the chickens. One Sunday morning my father went down to feed them and he found the shanty door wide open. He looked and saw that most of the chickens were missing. There was blood and feathers on the ground. He called my mother who was in the house. She looked out the window and he told her to come down. So, she went down to see what was wrong. They were yelling and asking each other how such a thing could happen.

Well after few weeks of losing more chickens, my father brought home a dog. This dog looked much larger then a shepherd. He was a mix breed and must have been half-wolf and shepherd. Now this dog was hard to handle. No one but my mother or I could feed him.

My father and Angelo Azzato, his brother-in-law, built a fence near the chicken coop to keep the crooks from taking things that did not belong to them. Now, this dog was a good watch dog. If any one went into the back yard, he would and try to jump over the six foot fence. He was a mean dog but a good watch dog. I was afraid of him. I would feed him most of the time and eventually the dog took a liking to me and my mother. Nobody else could get anywhere near the dog. This went on for a few weeks. After school, I would walk him in the yard with a heavy thick chain. The chain was about ten feet long and I would hold on tight trying to keep control of the dog. The dog was almost too strong for me to handle, but I managed.

This one day, my cousin, Joe Azzato, who we called Mulligan for short, came by to see me while I was in the backyard. I was walking the dog in the yard and Mulligan came into the backyard and stood on the top steps that entered the yard. It was about five steps high. He looked afraid of the dog. After a short time of standing on the landing, I told Mulligan to come down and be friendly with the dog. With some hesitation, Mulligan got about three feet away from the dog. At the same time, I held the dog and made the chain shorter. The dog barked at Mulligan and he was very afraid of the dog.

By instinct, Joe happened to put his right hand up to protect his face from the dog. The dog must have thought that Mulligan was going to swing at him and he attacked Joe. The half-wolf shepherd was too strong for me to handle and he got his teeth into Mulligan's leg and bit away furiously. I tried to get between the dog and Joe but the dog didn't stop. I yelled at Joe to run up the steps but he was screaming with pain and was frozen on the spot. I tried to push him away with my hip because my two hands were holding the chain and the dog's collar.

In the meantime, Joe's mother heard all the yelling. She was on the second floor and she started to scream, which only made the dog fight harder. Another few seconds went by and I was able to push Joe near the steps away from harm's way. At last, Joe was safe and taken to the drugstore by his mother to wait for the ambulance to take care of his leg. In the meantime, I put the dog back inside the fence. Then, I ran around the block and hid. I stayed away a good three to four hours. When it got dark, I managed to get upstairs into my apartment. Everyone was looking for me. Joe's father wanted talk to me. He wasn't the easiest guy in the world to get along with so I avoided seeing him until the next day.

In the meantime, the druggist told my mother she had to take dog to the ASPCA to have the dog checked for rabies. So, the next day, my mother and I went on foot about two miles or more to the ASPCA. For me, it was like going to hell and back holding that fierce dog close to me. I was still shaken up from the day before. He pulled me and lunged at anyone who got anywhere near him. This went on for about two hours for our entire walk to the ASPCA. When we got there, we had to wait another half hour before we could bring the dog into a room and have him checked for rabies. The ASCPA man gave me back the dog and so my mother and I had to go through the whole ordeal again to take the dog home. We couldn't take the dog in a cab or the trolley car, which is how we usually traveled in those days, because the dog was too dangerous.

We had to wait a few days for the test results to come in by mail. The dog was free of rabies and I thanked God for that good news. I saw Mulligan's father and he was happy that I saved Joe from any more injuries. Joe's leg healed but I don't know if he ever got over the fright from that attack. I still think of it after seventy years.

Louis had to work for everything he had. He had very few toys of his own and little opportunity for recreation. One summer, Louis and his older sister Kate saved the tops of Octagon soap boxes with coupons on them so they could cash them in for a pair of roller skates. They collected an entire shopping bag full of coupons. Louis and Kate couldn't wait to get a pair of roller skates. They had to take the train to go cash them in. When they arrived at the store, they were short a few soap boxes but the cashier let them use their train fare home to make up the difference. Kate and Louis walked home with their coveted pair of skates.

Louis finally tried the skates on and started to skate up and down their street. When his father spotted him, he was furious. He took the skates away from Louis and threw them down the street sewer. Apparently, Louis's mother and father didn't want them skating because it wore out the soles of their shoes and shoes were expensive to replace. John remembers getting in trouble for using a scooter for the same reason. Kate also remembered her mother watching her from the window of their home to make sure she didn't skate with her girlfriends.[16]

Angelina Lauria was a strong woman who raised 10 children. She rarely left the house and was always cooking and cleaning. She was protective of her children and looked out for them. When Angelina heard that Louis's teacher put tape on his mouth when he couldn't keep quiet in school, she returned to the school, gave the teacher a piece of her mind and proceeded to put tape on the teacher's mouth. She was a little woman around five feet tall, but wasn't someone you should cross. Once when a neighbor picked on Louis, she lifted the man up and threw him in a barrel.

Almost every Sunday, Angelina took her youngest boys down the street to Park Avenue and Walworth Street to their godmother's house. There was a lemon ice stand next door to her and the boys would beg their mother to buy them some. Lemon ice cost two cents for a paper cup of about five ounces. Many times she couldn't afford the six cents to buy the three boys lemon ice. Every so often their godmother would give the boys money for lemon ice, which would really upset their mother. She was not the type of person who willingly accepted handouts.

The Lauria children started working at a young age to help support the family. Louis's older sister Kate left school after a relative took her to work in the summertime. She worked in a sweatshop that made men's clothing. Her job was to take the loose threads off the finished garments. She made five dollars a week, which was a lot in those days, and figured for five dollars a week she could do the whole family's grocery shopping.[17] When she was supposed to go back to school and enter the seventh grade, she decided to go to work instead. She felt bad because she figured her mother could use the extra money. Their father was mad because he wanted Kate to become a teacher.[18] Louis's younger brother John also left school before the eighth grade to work for a plumber because the family was so big and money was scarce.

After Louis completed elementary school, he attended a special trade school to learn the barber trade and tailoring. From then on, he spent more time working than he did at school. When he was 13 years old, he was offered a full-time job at the American Almond Products factory that was right next to his home. His father knew the owner, Carl Peckler, from before Louis was born. He went to school one day a week and worked in the factory four days a week until he received his working papers.

He worked in the almond factory for 44 hours a week, receiving four hours overtime. He made $12.64 a week and gave ten dollars to his parents and kept $2.64 for himself. As a young teenager, Louis was a primary breadwinner for his family. American Almond Products sliced nuts for chocolate factories and bakeries. It used to make burnt almonds for ice cream pops, shipping its almonds to the companies that made the Bungalow Bar, Good Humor bar and many others. They also made macaroon paste and almond paste for several pastry stores.

Louis was very strong like his father and uncles, who were used to hard labor. Although Louis never grew taller than 5' 4", he was well known for his strength and massive hands. One of Louis's jobs was to unload 100-pound sacks of peanuts from trucks that brought all kinds of nuts up from the South.[19] His main job was to work the slicing machine. First, he soaked the nuts in very hot water, then took the soaked nuts and put them on a conveyer belt where the nuts dried up and their shells came off. Next, he put the nuts into a machine with a funnel and four very sharp blades that sliced the nuts. The sliced nuts fell onto another conveyer belt and then into a stainless steel tub, where they would dry out completely.

He also made the almond and macaroon paste. He would make flat pancakes and put them in a big oven, where they cooked to a golden brown color. Then he took them out and laid them on a stone to cool. After they cooled, he put them into a granulator machine. Then they were placed on wax paper lining in cardboard boxes and shipped to Bungalow Bar, Good Humor and other ice cream factories.

It was Louis's responsibility to maintain the machines in the factory and repair equipment. He greased and oiled machines, sharpened knives from the machines on grindstones and did other odd jobs around the plant. He was also responsible for training new men.[20]

In 1941, the family moved to Ozone Park, Queens, NY. Louis still worked for American Almond Products in Brooklyn. He had to take the train to work every day. He took the train on Crossbay Boulevard and Liberty Avenue to Broadway junction, went downstairs and took the Myrtle Avenue line to Nostrand Avenue and walked two and a half blocks to the factory.

Carl Peckler gave Louis two extra dollars a week and he was told not to tell anyone. Mr. Peckler liked the Lauria family. He helped the family out with jobs. Louis's two brothers Augie and John went to work at his resort in South Fallsburg, NY, when they were ten years old. They worked for two weeks when they were on school break.[21] In the summer of 1941, Louis's brother John worked on his farm, picking strawberries, weeding tomato plants, and helping in the kitchen. Mr. Peckler liked all the Lauria children, but he especially loved Louis.

Despite having very little, the boys found plenty of activities to entertain themselves. Louis, Augie, Frankie and John collected wood and anything that would burn starting in September for Election Day. They collected a massive amount and made a huge bonfire on their street on Election Day every year.

The family had pigeons that they kept on their roof in coops. Many families had pigeon coops at that time. On Louis's block there were two other families who were pigeon mongers. Almost every block in Brooklyn had pigeon mongers. This was one of Louis's hobbies. One night, Louis and his brothers climbed onto their neighbor's roof and stole hundreds of pigeons. They stuffed the pigeons and their eggs in a sack and started to climb down the outside of their neighbor's chimney. As they climbed down, the chimney collapsed. Luckily, no one was hurt. Their neighbor looked outside, but it was too dark to see the pile of rubble from the fallen chimney. The boys hid in the darkness until it was safe to return home.

Their younger brother Carmine snitched on them. When their mother found out, she opened up the door to their pigeon coop and let all their pigeons free as a punishment. The neighbor never found out that they had caused his chimney to collapse. That incident put such a scare into Louis, he began to turn around and get into less trouble. He felt lucky to be alive after falling off the roof unharmed when the chimney collapsed.

Growing up in Brooklyn, the Lauria children didn't have access to any playgrounds or baseball fields, so they played in the streets. One of the games they played was a game they called sewers. They would cut up an old bike tire into four-inch-long pieces. There were two teams and the object of the game was to hit the broken tire piece with a stick. The kids used old broom sticks or mop handles as their bats. If you were able to connect with the piece of tire, it went pretty far. When the children were lucky enough to get a ball, they would use it instead.

There was a spiteful old man who lived on their street and if the ball or tire piece landed on his sidewalk, he would take it and cut it up into pieces. One day, a ball landed on the old man's sidewalk and he grabbed the ball. Louis and his friend Frankie ran over and tried to take the ball back before he could cut it up. The man retaliated and hit them with a stick. Louis's mother was watching from her window and she came outside and ran across the street. She hit him over the head with his own stick and said, "if you ever hit any of my kids again I will kill you." He walked with a limp on one side and she said he would be limping on both sides

Summer of 1942, Louis and all his siblings, except oldest brother Ralph, at Howard Beach. Carmine lying down in the front. Left to right in laps: Rosie, with her feet on Carmine, and Danny. Sitting: Kate, Al, Matt, neighbor Mary, Augie and neighbor Rosie. Back row: Anthony, Louis and John (yelling).

if he tried anything again. He never bothered any of the children after that.

Louis fondly remembered growing up on Walworth Street and playing stickball in the street. He also loved going to Howard Beach with his family. They would rent a rowboat and he would go fishing or crabbing with his brothers. Although his childhood was filled with hard work, he was close with his brothers and cousins. The neighborhood was a tightly knit community made up of people from their parents' hometown, Marsico Nuovo. Everyone looked out for one another. Louis was loved and respected by all his siblings. They looked up to Louis and he made friends easily.

When he was drafted, his family and friends prayed for his safety constantly. Louis worried about leaving his family because he was one of their primary bread winners. On his draft record, he had to list his mother and siblings as his dependents.

After basic training, Louis was allowed to go home for a few weeks in the fall of 1943. He met his future wife Olga through a friend from the service whose sister was friends with Olga. Olga was dating somebody else at the time but quickly broke things off with her younger boyfriend after she met Louis. They dated for several weeks before he had to leave again and eventually went overseas to war.

Throughout the war they wrote each other letters. Louis sent her pictures from Europe with love notes on the back such as: "Olga I love you darling, Lou," "It's me to you. Love and kisses always" and "Be good darling, love Lou." Once during the war, Olga and her friend went to Manhattan, where she made a record of herself singing and sent it oversees to Louis. He used to joke that he should have played the record for the Germans and they would have surrendered willingly because of the torture.

Growing up as an older brother in a family of 12, Louis learned to put the welfare of others before his own. He was loyal and devoted to his family and friends. Louis had a remarkable work ethic. Everyone could always count on him to finish a job and to do it well. He started working at a very young age and wasn't resentful of his parents or his lot in life. He appreciated the little things and valued what mattered most. He never obtained fortune or fame, but to those who were lucky enough to have known him, he was truly a great man.

I
Going into the Army

I remember December 7, 1941, all too well. I was 17 years old. I lived in Ozone Park, NY, and it was three in the afternoon on a beautiful sunny day. My future brother-in-law came to our house and told me that Japan had attacked Pearl Harbor. At the time, I didn't know where that was, but I quickly learned. We listened to Gabriel Heater, the news commentator, describe how the U.S. was losing island after island in the Pacific. I had been positive that the U.S. would easily defeat Japan. I had underestimated the Japanese. I thought that by the time I was old enough to be drafted the conflict would have ended, but in less than a year and a half, I would be overseas.

In March of 1943, I received my greeting from Uncle Sam. The letter stated, "You are invited to the induction center at Penn Station on the 25th of March." So, I accepted the invitation. When I arrived at Penn Station, I had to have my physical, which took the entire day, and I passed. Once that was over, I was told to report to my induction center on the first of April, 1943.

I couldn't believe that the draft had called me so soon. The draft age at that time was 21. In December of 1942, my brother was called at the age of twenty and a half. I was in shock. Then the draft age dropped from 21 to 18, and so I was called. On the first of April, I had to go to my induction center at eight o'clock with all the boys who lived in my area. The induction center was about a half mile from my home.

My mother's uncle, Angelo Azzato, picked us up. He lived in Brooklyn and had a car. I said goodbye to all my younger brothers and then Angelo drove us to the American Legion hall which was next door to the draft board. When it was time to get on the bus, my father said goodbye with tears in his eyes. My mother held me tight and kissed me and told me to take hold of her rosary beads and to always pray with them. I gave her a big hug, kissed her and then got on the bus when my name was called to go to the Long Island Railroad Station. I was sad leaving my mother and father behind, not knowing what awaited me.

From there, we traveled to Penn Station in Manhattan and I grieved all the way there. Here many other citizens also started their journey to become soldiers. There was another roll call in Penn Station before we were on our way to Camp Upton in Long Island, which was about 70 miles from my home in Ozone Park, NY. Remember, in those days, Long Island was nothing but farmland and wooded area. Not many people had automobiles. On the way to camp Upton, I was not very happy about leaving home.

My mother and father had seven other kids at home. Besides my father, who was a street cleaner with the sanitation department in Brooklyn, my sister Kate was the only one working at the time. My father was nearly 56 years old. My brother Matty and I weren't going to be home to help support the household. This gave me something to worry about. My oldest brother Ralph was married, had two kids of his own and worked at the shipyard with the war effort. He worried about being drafted. He was called a few times but turned down because he had two kids and he was already working at the shipyard. He was lucky the draft board didn't accept him and he got a deferment.

It was a sad ride for me to Camp Upton. Lots of the boys joked around and had fun but I wasn't in the mood. The train ride was three to four hours long because the train had to stop for other trains to go first. The trip seemed so long. Two of my relatives lived a few blocks away from me and left on the same train as I did. After we got to Camp Upton, I didn't see them again until the war's end.

Getting off the train was just as bad. We had to wait for our names to be called. After that, we marched to our designated barracks. We got a bunk and then were sent out again for haircuts. The GIs got haircuts almost down to the skin of our heads.

After that was over, we lined up again to get our shots for the day. As we waited for our shots, some GIs, who had been there for a few days, yelled out to us, "Watch out for the hook!" They put a scare into us and made us sweat this thing out even more. When we walked to the aid station, they yelled out again for us to watch out for the hook, trying to frighten us.

After we received our shots, we had to go to the supply house barracks with all the army wear. At 5' 4", I had a hard time finding the right size and half my garments were too large. The shoes were too big, the pants were too long, and the shirt was too baggy. I had a rough time getting my clothes.

Next was supper time. We brought our clothes back to the barracks and fell out for chow. That mess hall really was a mess. We had to get used to all the noise, with many of the guys yelling one thing or another. If you were at the end of the table, you had to yell out for coffee or pancakes or whatever was on the menu. You couldn't stop the food from getting to you first, that was a no-no. It had to go straight to the man who asked for it. That was part of the new manners we were taught. Each day we learned something new.

After supper, we went back to our barracks and picked bunks with footlockers in front of the beds. Later on, the sergeant came in to show us how to make up our beds, how to hang our clothes, where to put our shoes and how to polish them. The bed had to be made very neatly. The instructor sergeant told us how things had to be done. We were mostly treated like nincompoops at the beginning and had to be alert at all times. The days seemed long, and we were exhausted by the workout. We were on the go the whole day. Nighttime arrived and we got ready to go to sleep. Lights went out at eleven o'clock, the bugler blew taps and all was done.

I got into my bunk and held the rosary beads that my mother had given me when I got on the bus, rested my head on the pillow and started to cry. I knew that some bad things were going to happen to me in the horrible war. I cried myself to sleep. It must have been about midnight.

Time went so fast when you slept. The next thing I heard was the bugle call. It was about 4:30 in the morning. The sergeant came into the barracks and yelled, "Let's go, everybody up! Drop your cocks and put on your socks!" This was really something to go through. You were dead tired when you went to bed, and the next thing you know you are getting up, half scared to death with all the noise.

The sergeant blew whistle. GIs were whining and complaining because we had to get up so early. The sergeant kept blowing his whistle and yelled, "Let's go, everybody up!" We all rushed to get dressed and fell out to be counted in the field. We lined up and the sergeant called out our names. We answered "here" when we were called. We were told to fall out and go to chow, which was breakfast. After we ate, we were told to go to our barracks and wash up. There we waited for the whistle to blow for us to fall out again.

After the whistle blew, we fell out and waited. It was still dark outside, not yet springtime. It was the first week of April and six o'clock in the morning. We stood in a line and the sergeant came close to us, pointing his finger, and said, "you, you, you, and you," and told a few others to step forward. He had one of the corporals march us to a place about 400 yards away. After we got there, he told us to pick up a wheelbarrow and a shovel, and pointed to a coal pile. There were five or six piles of coal over 20 feet high. The corporal told us to shovel the coal into the wheelbarrows, fill them up and wheel them about a hundred yards to the furnace. We brought the coal into the plant and dumped it near the large furnace. There were about five large furnaces inside that plant, making heat and hot water for the entire Camp Upton.

So, I and about nine others were picked for this work. We started about seven in the morning and only stopped for chow. Afterward we went right back to the coal pile and pushed the wheelbarrows back and forth until about five o'clock. At times, we had to shovel the coal into the furnace. It all depended on how much coal was needed in the furnace.

When we were told to stop, we were physically bushed. We waited for a truck to bring all of us back to the barracks. We had to clean up and get to the mess hall before all the chow was gone, although I was so tired I didn't have the strength to eat. After we had chow, we went back to the barracks and I got in bed to get some sleep. My legs and arms were in a lot of pain. After a short time, I awoke and sat up on my bunk. I kept thinking what fate awaited me. It was almost time for lights out. The bugle blew and all the lights went out. I fell asleep again within seconds.

The next thing I heard was the bugle with the sound of reveille. It was about four in the morning. We went through the same thing as the day before; we fell out to be counted and then went over to the mess hall for breakfast. After that was over with, we went back to the barracks and waited for whatever was next. We waited awhile and then here came the sergeant, blowing his whistle. We all went out and lined up. The sergeant did the same thing as he had yesterday morning. It was still dark outside. He came closer to us and did the finger point, "you, you, and you." And so on. I happened to be picked again for the same detail. This went on for a few days in a row. I was exhausted doing the same detail and I was used to hard work, having worked in a peanut factory since I was 13 years old.

After four days of the same coal detail, I heard one of the GIs tell another GI that if we stayed here for 21 days we would get a pass to go home for the weekend. I didn't like that idea at all. I wanted to get out of Upton as soon as possible and go to another camp. Everything was the same at Upton and the next morning we would be up and at it again.

Every morning after breakfast, we waited for roll call. If your name wasn't called, you stayed for a detail. If your name was called, you were shipped out to a new camp. Well, after a few days at camp, shoveling coal, my name was finally called. I was going to get to ship out of this hellhole. My name was called with those of a few hundred other GIs. We were to get ready to get on a train and head for some unknown destination, where we would get 13 weeks of basic training.

My official army photograph taken during basic training.

We GIs didn't know where we were going, but we would know when we got there. The rule was tight lips. The train trip took at least two days. I remember we used to call the troop trains cattle trains. We slept on the wooden seats, which were not too comfortable, and we wore the same clothes. It was sweaty and smelly going down to North Carolina, where it was much warmer. It was some five hundred miles from New York and a big temperature change. It went from 35 degrees to 75 degrees in two days. Instead of needing longjohns, I wore shorts, but I didn't mind the heat. I never did like winter. I guess I cheated the weather by being shipped to the South.

We arrived at Fort Bragg, NC, and were assigned to an artillery outfit. There were a few artillery battalions in that camp, but it was the camp with the 82nd Airborne Division. Basic training went on for 13 weeks. I did make new friends. It was a tough deal, but I made it through with flying colors.

While I was stationed in Fort Bragg, my brother Matty had gotten a weekend pass from the camp where he was stationed at Fort Fisher. It was located just outside of Wilmington, NC, about 100 miles from Fort Bragg. Matty wrote me a letter telling me he was going to visit me as soon as he got a weekend pass. His dream came true and about two weeks later he arrived at Fort Bragg. I was very happy to see him. My older brother was with me and I had tears in my eyes. I was relieved that I didn't have to do guard duty that weekend so I could spend more time with him. I was still doing my basic training and there was no way in the world that they would have let me out of my shift.

Matty and I went to the P.X. and had a photo taken which came out pretty good. We had the photo sent home to our parents and it hung in our parents' home for over 62 years. After my brother left Fort Bragg, I had the opportunity to visit him. We did this several more times until I finally finished my basic training. I did not see Matty again until the war ended.

You should have some idea how things were done in the army. Getting up at reveille, the GIs tried to get to the toilet or what the army calls the latrine. Our barracks had 60 or more GIs and everyone had to get to the latrine fast to wash up and do whatever necessary to look clean. Our khakis had to be spic-and-span. We had to be cleanly shaven, but I didn't have to shave. Sometimes we would fall out and line up shoulder to shoulder. The sergeant told us to put our heads down and look for cigarette butts and paper wrappers or anything that did not belong on the ground. There were many little details that had to be taken care of throughout the day.

My barrack sergeant, Sergeant Trota, took a liking to me. He always called me "Shorty." He always said good things about me when we were doing close order drills. I was very good at it. As a matter of fact, at the end of our 13 weeks there, we had a contest on the large field and a few generals came to see us drill. My barracks won the contest and the general gave our sergeant a lot of praise because he was a good instructor and a good mess sergeant.

Near the end of my basic training, Sergeant Trota wanted me to stay at Fort Bragg and be a cook. He said, "Shorty, I want to send you to cooking school to become a cook. You will not have to go overseas in combat." He told me to think it over and let him know, but I didn't want to stay. He approached me a few times about it. I told him that I wanted to stay with the men. I knew he was sad when I turned him down. He said, "Shorty, you will be sorry." I told him I appreciated the thought.

Sergeant Trota was a fine man. When we left the camp, he said goodbye and good luck to me.

The boys that were leaving boarded a cattle train again and we headed for Camp Rucker in Alabama as replacements in the 35th Division. We were attached to an artillery unit with 155 howitzers. They were bigger guns than the 105 that we trained on in Fort Bragg. Most of the GIs in my new barracks were American Indians and they were a wild bunch of guns. If they got drunk on a weekend, look out, they really went wild.

I remember one day we went to the firing range to see how the 240 cannons were used. When we arrived and saw the size of the cannon, I was afraid to see it fired. The shell was so large it had to be put into the gun with the help of a crane. They warned us to keep our mouths open and not put our hands over our ears when it fired. Well, we did just that. I was scared like all the rest while we waited for the gun to be fired. This thing was like the Big Bertha cannon used by the Germans in World War I. We heard it go off and, boy, it was a big bang. Our ears rang for some time after and I was happy they only fired one round.

I spent about two months training at Camp Rucker. Then we were shipped out to Fort Meade, in Maryland. At Fort Meade, we waited for standby. I knew we were getting some new equipment. There was a lot of talk that we were getting ready to go overseas. I believe we stayed at Fort Meade for about two weeks and then we were on the move again.

We were back on the cattle train, not knowing where we were going until we arrived there. Our destination turned out to be Camp Myles Standish, near the Boston harbor. Now we were sure that we were going overseas. I believe I stayed there just a day or two. I don't remember how much time I stayed in that camp. I am sure I did some K.P. (kitchen patrol) there. Boston harbor was set up for embarkation; soon we would board a vessel and go overseas.

II
Overseas

We boarded the ship *Mauritania*, which was built in 1913, a year before World War I. At that time, it was the third-largest ship in the world. When we boarded, we were given a place down inside the ship. Most of us were assigned a hammock to sleep in. After that, we got to go up on deck and look around. It was a scary thing looking down; it was a long way to the water. I was sure it took the crew a good day and a half before we set sail. There were two thousand nurses, two thousand officers and ten thousand enlisted men, not counting the crew members. When we set sail, the ship moved out and a few tugboats guided it out of the harbor. It was October 9, 1943, when we departed for England. After it got out in the open waters, the tugboats left and the *Mauritania* embarked on its own.

Looking up, we saw a Catalina plane overhead. It kept an eye on the ship. I was happy to see that someone was looking out for us. The deck was full of GIs and there was barely any room to stand. The PBY Catalina passed over the ship looking for any sign of enemy subs. It stayed with us for a few hours and then went back to its base. I wished it would have stayed longer.

After that, most of us went back down to our assigned hammocks. I put some of my equipment nearby, got into my hammock, and tried to get some shut-eye. The rolling of the ship started to make me dizzy. I got myself down from the hammock, which was about the second from the bottom. I was lucky because the hammocks went maybe five high. I started to head up to the top deck because of the dizziness. I thought I was going to throw up. On the top deck, there was a shore patrol, or what we in the army call an M.P. He wanted to know what I was going out on deck for. I told him that I wanted to get some fresh air because I was feeling dizzy. He said, "Okay, but no smoking on deck," and I told him that I didn't smoke. I stayed for about five minutes and went back to the

M.P. to let him know I was going down to get my coat. It was very cold out on deck. I went and put my coat on and then went back upstairs. I was only about three stairs down. Going down three stairs wasn't too much to walk. I did that real fast because I did not want to get dizzy again. I got back on deck and tried to keep warm, but the wind was really blowing at a good speed and the spray from the strong waves got me wet. I didn't know what to do. Stay on deck and feel much better or go down to where the hammock was?

I decided to stay up on the deck. I looked around for a cozy place to hide and stay warm. I tried to find a dark corner on the ship to hide so the M.P. wouldn't chase me back down below. I began to walk the deck and the wind and ocean spray got me all wet. Finally, I saw this large roll of heavy rope which was neatly rolled and large enough for me to get inside of, like a fetus. The large rope was for holding the ship to the port holder when it was docked and towing the ship. It was called a hawser. I put myself inside of it and made my heavy coat cover me as much as possible to stay warm.

I must remind those who read this story that it was October and we were sailing in the North Atlantic. The temperature dropped quite a bit. The waves grew larger and almost came on the deck. Every time the bow dipped into the waves, the stern of the ship came out of the water and the propeller would vibrate. For seven days on the *Mauritania*, I slept inside that hawser. Each day, I had to get up and out of it before daybreak. This way the M.P. didn't see me sleeping inside it.

The conditions were not the best, but I did manage to stay alive each night. It got even colder as time went on, but I made it. Down in the lower deck all the GIs were so crowded together that there was no room for anything else. We were like sardines in a can. It was noisy and chaotic down there, so I decided to stay up in the cold with the freezing water spray.

I believe it was on the seventh day out that we spotted a German plane about twenty thousand feet above the ship. Within seconds, the British gunmen opened up on the plane and shot a few rounds at it. The German plane took off. I guess it was just doing some reconnaissance. The next day there were around five thousand guys on deck that day, anticipating our arrival in England. A few hours later, somebody said they saw land ahead. All of us GIs smiled. We were lucky to have made it without getting hit by a German torpedo sub. We were all happy to see England ahead.

We landed in Wales, UK, on October 17, 1943, after 8 days at sea. After we left the ship, they took us to a place called Litchfield, England. The camp had a lot of barracks and we were assigned to one of them. England always looked so foggy.

I remember that I had gotten a bad case of ringworm and it spread all around my bottom and my privates. I didn't want to go on sick call because I was afraid they would put me on quarantine and separate me from the boys that I'd been with since the training. I went to the P.X. store looking for some medication to put on the ringworm area. I didn't find anything to use, so I picked up a can of lighter fluid and went back to my barrack.

Late that night, I went to the latrine and got some toilet paper, wet the paper with lighter fluid and put it on the infected part of my body. Then I went back to my bunk. I was on the top bunk. Oh God, I was sorry I did what I did. The lighter fluid burned so bad, I wanted to cry and run down to the latrine to douse water on the infected area of my body, but I decided to suffer the pain for about three or four hours. As a matter of fact, I fell asleep with the pain because I was so exhausted. I had had very little sleep the last seven days, trying to sleep in the hawser of the ship.

Reveille came so fast I couldn't believe that I had had three hours of sleep. I heard the bugle call and was up and at it again. I washed up and assembled outside. After roll call we were told to go to chow. We only stayed in England for a few days. Luckily, my ringworm healed up, but I wouldn't recommend using lighter fluid to anybody.

Ireland

One or two days later we were on our way to Scotland and from there we crossed over the Irish Sea. Then we got on a boat to Belfast, Ireland. It wasn't a large ship. It was more like a ferry. I remember there were about three hundred men on board. We found out later that we were replacements for the Fifth Infantry Division.

Most of the guys were separated but some of the men I was with from the States also went into my division. I was assigned to the Cannon Company of the 11th Infantry Regiment with some 15 other boys from Fort Bragg. While in Ireland, we trained for ten months with the Fifth Infantry Division.

The Fifth Division was stationed in Iceland for two years before I was part of it. They trained in Iceland, and then were shipped off to Ireland, where I joined up with them in October of 1943. The division was a hardened outfit but had had no combat experience since World War I. In 1941, it was sent to Iceland because the U.S. had to stop the Germans from taking over that island. It was not an easy task. The outfit had to put up with freezing weather. There were only a few hours of daylight during the winter months. Thirty of the new replacements, including myself, joined the Fifth Division in late October 1943 and began training with them.

We had to put up with the rainy conditions in Ireland, which lasted for the entire ten-month training period. A new man like me had to listen to the other men's stories about what they endured during training in Iceland. This was the new men's punishment for not spending time in Iceland. For ten months, all we would hear was, "You guys don't know how lucky you are." They constantly told us how fortunate we were to have missed out on the extra time in Iceland that they had to experience. They told us about the towns Reykjavik and Keflavik. We young guys knew that Iceland had been a bad ordeal but it didn't make any sense to take it out on us. We were not of age when they were drafted so I let it go at that.

While in Ireland, we formed a cannon company. Each regiment had a cannon company. We trained on the new 57-millimeter weapons and with the 105 howitzer cannon, which was small and powerful and had a range of five miles. This replaced the 37-millimeter anti–tank gun, which was obsolete. We used it as a mortar and for direct fire on tanks. It was an anti–tank weapon. The company trained in Ireland on the new cannon for ten months, preparing for the land invasion which we had no idea where or when would take place.

Most of us GIs talked about the invasion and knew it was coming. We guessed about where it would be. During the last week of May, our outfit received an order to board a ship in Belfast. Some of the older men in the outfit had a notion that we were going to be shipped home. They had been overseas for three years and longed to return to their families. They guessed wrong. The 11th Infantry left for port, but I was left behind to guard the camp with some other GIs. We had to wait until another outfit arrived to take over our previous position.

Guard duty consisted of two hours on and four hours off. The first two nights were cold, windy, and rainy. This type of weather was nothing new to us. It rained almost every day while we were in Ireland, with the exception of about 15 decent days. When I walked my post, I could almost fall asleep while standing up. Many thoughts would go through my mind. I always disliked guard duty, but it had to be done, so we did it. It was unpleasant, but not nearly as bad as landing in Normandy. Later, I wished I could be back in Ireland doing guard duty again. There is nothing in this world that can make you feel happy about combat. It sure was hell, and I would have given anything to return to the innocent times of the past.

While I was walking post in Ireland, on my second night, some time around two bells, I heard a GI on the night post yell out to me that the invasion had begun in Normandy. I didn't know what to say. I got down on one knee, made the sign of the cross and said, "Lord, look over them." I thanked God and asked for his help. I said the Lord's Prayer from the moment I heard the news of the invasion until it was time to finish guard duty. I was afraid and happy at the same time. I knew it would not be long until I saw the battle too. The next day around eleven o'clock, I heard troops coming into the camp. It was our unit returning. I wondered why, but never did find out why they returned. It could have been a decoy maneuver. A week and a half went by and the landing turned out to be a success.

III
Normandy

On July 7, 1944, our unit set sail for Normandy. Our unit boarded a Liberty ship named the *Excelsior* which was a cargo transport ship. "My God," I said to myself, "We wouldn't make it across the channel on that ship." Not so much because the ship was small, but because it looked very sinkable. I believe the entire 11th Infantry Regiment was on it. I am sure we started in the daylight and that every one of us looked scared. We all tried not to show it though. Every so often, one of the boys brought up the subject and said something like, "Gee, I wonder how things will get over in France." The trip was about two days long. All of us had our life jackets on. I never took my jacket off because I couldn't swim. I slept with the darn thing on. We kept our gas masks on us at all times too.

All of us GIs had C- or K-rations to eat day in and day out. When it was time to eat, we put our canned ration into a garbage can with everyone else's can. The garbage can was half full of water and it had a heating unit. Once the water heated up, we formed a line at the garbage can and somebody in charge gave us a can. It was like a potluck. Our bread was a biscuit. We also received a hot canteen cup of water that we put our coffee powder in, without milk.

One day, I stopped a merchant marine that was part of the kitchen staff. The smell of real coffee that drifted from the kitchen drove me crazy. I used to drink anywhere from 15 to 20 cups of coffee a day before going to the war. The aroma of the real coffee tortured me. I begged the merchant marine, who I was friendly with, for some coffee. He told me he would give me a cup when the time was right.

I waited for almost the whole trip for some coffee, and he didn't come through. My mind was not on the war but on getting that cup of coffee. On the last day of our trip, when I could see the shoreline of Normandy, I asked him again for some coffee. He realized that time was getting short, so he told me to wait until most of the GIs had left the mess hall. Once the area cleared, he told me to get my canteen cup ready and hurry over to the mess hall, which was an open area on the ship deck. I quickly walked over and he took my cup and filled it to the brim. I went over to the side of the ship and started to drink it as fast as I could. Oh boy, it was a heavenly taste. Now that was real coffee! After drinking it, I went back and thanked him. He said it was no big deal and then wished me all the luck in the world. That cup of coffee has stayed on my mind for over 56 years.

After the drink of coffee, we all stood in a group with the merchant marine, looking at the thousands of ships in and around the shoreline. We could not believe our eyes—oh, what a mess. You would have to have been there to believe it. Then the merchant marine told all of us what had happened, on D-day. He explained how so many men had gotten killed and how the men who were manning the machine guns on the ships were so nervous that they shot our own planes and sometimes hit them with machine-gun fire. It was complete madness to hear that our men were shooting our own men. I guess there was so much confusion that everyone was an enemy.

After a short time of listening to the merchant marine, I got nervous and walked away, looking out to the beach where medics were putting dead soldiers in rows in the sand. Then I looked over by the cliff which was about a half mile to my left. This is where the rangers climbed the cliff called Point de Hoc.

Things were looking sad. Everyone was talking about something. About an hour went by and the message came over the loudspeaker that all personnel would be disembarking the cargo ship and we should proceed in an orderly way. The loudspeaker called out each company. A Company went first, then B Company and so on. Cannon Company was almost last. This went on for five hours. It was July 9, 1944.

We had to go down the ropes along the side of the ship with a backpack, gas mask, life jacket, and a shovel or spade for digging foxholes. Before I went over the side, I thanked the merchant marine again and he put his arms around me in a big hug. He wished me all of God's help. I then said the same thing to him, telling him to get home safe. I turned my head so that he wouldn't see the tears running down my face.

Going down the ropes on the side of the ship was a scary thing to do. The ship was moving from side to side and when you looked down, thinking the ropes were in the landing craft, you would be looking straight into the water. I hung on for dear life. Every one of us had the same feeling of gratefulness because we didn't disembark the first day of the invasion. It was scary the whole way down. Getting your footing on each rung was a job in itself. Well, after climbing down the unsteady ropes, we were now in the landing craft.

The time had come to go ashore. We started to move after the commanding officer gave orders to take off. When we get to the shore, the craft came to a fast stop. The ramp was lowered and the men were told to jump off. To my surprise, the water was about two and a half feet high. I waded through the water with difficulty with my backpack, rifle, ammo, hand grenade, gas mask and life jacket. We waded a good fifty yards.

Getting on shore, we all bunched up, wet to the bone, waiting for new orders. It was starting to get dark. We were beginning to get cold from our wet clothes. Finally, we got orders from our captain to go over the bluff and walk five miles to a new area. The company trucks picked up our duffel bags and we walked. It started to get dark about 19 bells. We stopped in this small town and were told to rest for the night.

We took out our K-rations and hot water for coffee. Then we walked guard duty. I slept in my wet clothes that night. During the night, I could hear the noises of combat in the distance. We heard the sound of artillery and machine-gun fire. The sky lit up as far as the eye could see. There was an air raid in Cherbourg, where the British and Americans bombed the Germans. We could see the tracers from the German anti-aircraft guns in the sky. There must have been thousands of tracers. This went on for most of the night. We talked and watched and wondered how we were going to cope with all of the horror. It never did get easy. Every day that went by, the situation worsened.

The next morning came too soon and the horror was all around us. The outfit was told to move to a new position. I remember we went through the town of St. Mere, where everything looked so gloomy. We could see where airborne troops and gliders had cracked up, trying to land on the night of the invasion. Most of my company was getting shaky. I know I was because I found it hard to swallow.

As we got closer to the front line, shells started to come in on us. I couldn't tell what was coming in or going out. Our outfit was green and had not seen any combat. Our unit took over for the First Division, which was called the Big Red One. We learned how to fight in combat. It took us a few days to get our bearings and understand what was going on. We stayed on the front line for ten days until a British unit replaced us. During those ten days, our unit got a taste of blood. I noticed the absence of some of the GIs in our company. I asked a friend where they went, and he told me that they had gotten hit and died. I couldn't believe how many men could lose their lives in such a short amount of time.

As time progressed, the green troops became hardened to the horrors of war. We moved again to a new position. Our unit used a give-and-take strategy that depended on losing some men in exchange of a greater loss of enemy troops. The enemy constantly entered our position. I remember the first day in combat when one of our men said, "Hey Lauria, there are five Krauts dead over there in a foxhole." He wanted me to go over to see them, but I refused. I was sure that I would see plenty of dead Germans by the end of the war. I knew that there would be many horrific scenes awaiting me. Our outfit would encounter all sorts of danger.

I believe in the ten months of combat I must have dug over five hundred foxholes. There were days I may have made three, four, or five, depending on how much ground we would advance. We were on the move and at times we would advance a few hundred yards or a mile or two. We would come to a halt, dig in and then wait for a new attack. The fighting was hard going.

We dodged mortar, artillery, small-arms and machine-gun fire most of the day. We sweat and prayed for an end to this mess and by the time it was nightfall we were completely exhausted, too tired to eat a K-ration or C-ration. If we halted for the night, we could fall asleep on a picket fence. If the night had no activity, we might get away with just two hours of guard duty for the night. The nights went so fast when we needed sleep.

Radio Wire Crew

I was a part of the Cannon Company, which consisted of three platoons with two howitzer cannons for each platoon. We supported the 11th Infantry Regiment. I was part of the radio wirer section. My duty was to run a wire to each cannon. Each section had a field telephone and I hooked my wire to their telephone. I had to haul around a large field phone and a reel of wire. Sometimes, we used a walkie talkie and at others we used a land line. In city fighting it was hard to receive good communication, so it was wise to bring along a land line. I also served as a spotter. I would go up with the riflemen and if we were pinned down, I would call in fire missions and get a fix in the area. The air force would come in and knock down the German position.

Members of the Cannon Company and radio wire crew. Clockwise from top left: Pop Rocky, Owen Stanley, Henry Bogart, Steve Maleevich, Turner, Harry Bogart, Grant Campbell, me, and Len Tredanari.

Our captain was Harry H. Smith, and he proved to be one great officer. In the radio and wirer section, Sergeant White was the recon sergeant and Sergeant Turner was my radio wirer sergeant. The radio wirer section consisted of about eight men at one time, generally including Pfc. Tredanari, Pfc. Malsavich, Cpl. Godzisz, Cpl. Hicks, Pfc. Campell, Pfc. Chester Pilip, Pfc. Pop Rocky, switchboard operator Pfc. Bogart, Pfc. Dutkiewitze, Pfc. Turner, and myself, Pfc. Lauria.

When the riflemen got pinned down, it was up to the radio man and his partner to get them out of harm's way. If our officer, who was with us at all times at the observation post, saw a machine gun, he would give us orders to call command post (C.P.) for a fire mission. We would tell the number one gun to put one shell of smoke in the cannon. We had to have an aiming point to go by. At command post, they would have to use a map to calculate about how far away the target was. If we requested a round of smoke in the daytime, it was easy to spot. If it was not close enough we would call back and tell them that it was a hundred or so yards too long. Then the C.P. would adjust the cannons. It may be a hundred yards to the left or right, so we would tell him this and they would calculate the new mission. They might also have to open up the shell and take out a powder bag to change the elevation.

My buddies. Left to right: Owen Stanley, Len Tredanari, Steve Maleevich, Harry Bogart and Grant Campbell.

I laugh at some of the movies they make in which a radio operator calls a fire mission and it just seems to get done so fast. This is all bull. The officer with us gave us the command to say and we repeated it to the command post over the radio or telephone. There were times when we were lucky and the shell would land and explode a few hundred yards away, it could be seen, a correction could be made and we could bring the shell almost within a yard of from the enemy. Then we would tell the C.P. to have the other five guns converge on the number one gun and then all six guns would fire. If all went well, we had a chance of getting the enemy much faster.

Day in and day out I was on the radio or telephone giving fire missions. Every day the code words were changed so the enemy could not tell what was going on and who was speaking. When we changed the code, it was only from a 0 to 3. We were given three letters that would be the radio wire code. After this was repeated, we gave the platoon number and code. My code call was dusty black two because I was in the second platoon. My officer would be number two. The company had three platoons, each with two howitzer cannons, which gave us six guns in the company. So, we had dusty black one, dusty black two, and dusty black three. Our captain's call code was dusty black six and the executive officer's was dusty black five, the higher the officer the higher the number.

The code on our radio was something like this: able baker Charlie two to able baker Charlie six. We would get three different letters in the alphabet and use them. For instance, if I had to use the letter C, I would say coyote, B would be baker and Z would be zebra. I used the number of the officer I wanted. For instance, if I wanted to get the third platoon I would call on the radio: coyote baker zebra two to coyote baker zebra three. He would then know we wanted to speak to him and he would get on the radio or phone. So now you get the picture, I hope.

The executive officer was in charge of the command post sending out the fire missions. He would deliver the command to fire the guns. If I was at the command post, he would tell me and I would repeat it to the guns.

Incidentally, there were only six howitzers to the regiment, two for each battalion. There were times when just one of our platoons would have to support a battalion during an attack and there would be other times when two platoons would go out in support of the battalion. It depended on the situation. The howitzer played a big part in the regiment. It was a short gun but it played a big part in knocking out enemy positions, troop movement, and tank attacks.

When we prepared to cross a river, the guns would keep a constant barrage on the far side of the river, knocking out enemy guns and giving our foot soldiers the necessary time to get across the river. We had the cannons fire at will constantly. If a target was spotted, we would change the reading to knock out whatever it was that was firing on our troops.

I wondered how I got stuck in the radio and wire section. Today it still baffles me. In Ireland, at the beginning, I was a cannoneer in the Cannon Company. I didn't like being a cannoneer. Training was hard going. Just moving the gun into position was difficult. It was a small cannon, but still hard to maneuver, plus, after getting it into position, you had to dig a ditch around most of the cannon, which I didn't like doing. I did like the fact that everyone got to go from being a gunner to pulling the lanyard and setting the shell or changing the fuse if needed. Still, I didn't like being a cannoneer.

One day during training which we called cannoneer hop, one GI named Wineburger, who had already been stationed for two years in Ireland, said to us, "Why don't you boys sign up for the radio and wirer section?" He told us we would have it much easier. He explained that the radio wirer would have to walk out on the field with the sergeant and take along the equipment. He told us that after a few hours of work, we

Me by one of the howitzer cannons.

could lie under some trees and pass the time of day. I thought at the time that he as much wiser than the rest of us. It turned out to be easy, but not that easy. When we trained firing the cannon, we had a better idea of what we would encounter during combat. The only difference was that nobody was shooting back at us. Little did we know what was in store for us as wire men. Most of us who signed up for the wirer section were the new guys that had just joined up with the Fifth Division and didn't know any better. We were in Ireland for about a month before we signed up. Radio and wiring section trained day and night while the platoon trained every other day.

Our sergeant in Ireland was Sergeant Webb. He was a good sergeant and all of us in the platoon got along with him. Later on, he was taken out of the wire section and placed in one of the gun sections and Sergeant Turner took over the radio section. Even after the intense training, we still had no idea what was in store for us. Later, when we were in combat, we realized that Wineburger had tricked us. About a week or two before we went into combat, this doughboy, who spent two years in Iceland, said to all the boys in my section, "You guys are in for hell if we go into combat as radio and wiring men." He explained to all of us what our chances for survival would be. After he finished talking to my section, I got scared and turned green. I realized that I had made a big mistake.

It turned out that Winburger owned a bar back in Chicago and wanted to get back to the States. He told us that he was going to break his leg to get out of going into combat. A few days went by and I heard that Winburger was in the hospital with a broken leg. He had deliberately smashed his leg as hard as he could on some steps that were in front of one of our huts. Things turned out just as he planned. We never saw him again. I know he got away with it because he wrote a letter to his buddy, who was in my section during combat. My friend let everyone read the letter. Winburger said he was on his way home and he was wearing a leg brace. He broke his leg in several places. He went home and probably received a pension from the government. He certainly was a con-artist. There is no justice when you try to be legitimate. His choice made him a real coward despite his having the guts to break his leg at the time. I wonder how he made out at home. He may have told a lot of war stories to all of his friends in the bar.

Broken Teeth

On the second day of combat, I didn't know if the mortar or artillery shells were going or coming. My outfit was getting what was called battle inoculation, which give green troops a chance to get used to combat—something which I believe no one ever really gets used to. We learned fast which shells belonged to the enemy and which ones were our own.

I was in my foxhole, which was dug by the First Division men, and decided to get out and walk around for a few minutes. I only walked about ten feet away because I was afraid to go too far from my hole. Ten shells landed in the field we occupied. I jumped up and leaped for my foxhole. I was so scared. I dove headfirst and hit my mouth on the side of the hole, which was about four feet deep. I was so frantic that I didn't realize that I hurt myself in the process. I was too scared to feel any pain, although the impact loosened one of my front teeth.

After this accident, many months went by and there were never any signs of pain. I remember the weather had changed and it was starting to get cold. Winter approached. I bedded down for the night in a barn and slept with a camel-hair blanket. The night rang with gunfire and was full of shelling and machine-gun activity. A GI woke me up in the middle of the night with the news that we were moving forward to an observation post. I had terrible pain in front tooth. I tried to drink some coffee and eat some of my K-ration, but the pain was unbearable.

I asked my top sergeant to go to the rear and see a dentist. He gave me the okay. A jeep driver drove me to the aid station. From there, I mounted a much larger truck with around ten GIs who were also sick. We drove about 40 miles or so away from the front lines. I guess that was the closest dentist around. When we did arrive, it had to be about three in the afternoon. I jumped off the truck and went into a building that looked like a hospital. There were many GIs in clean olive drab. They were all doctors' aids. They stared at me with wide eyes as if I walked in slow motion and were shocked by my untidy appearance.

You must remember, I came from the front lines and not some fancy hotel. My shoes and pants were full of mud. My face looked just as bad. My face was swollen from pain and very dirty. I may have never looked worse. It was a good thing I didn't have to shave because it would have made me look even worse. One of the medics asked me what I was there for and where I came from. I tried to explain to him why I was there but my tongue hit against my tooth and made it too painful to talk. The medic called over a few more GI medics. They started to ask me the same thing about where I had come from. I tried so hard to explain to them that I was from the Fifth Division and was there to have the doctor look at my tooth.

They were amazed and kept analyzing me. They were bewildered and asked me many questions, wanting to know how I endured the life of an infantryman. I told them that I had worn the same clothes for over two months without a bath of any kind. I told the medic that this is what I had looked liked for months. They also asked how I slept with all of the shelling and gunfire. They wanted to know if I slept in a house. I explained to them that I slept in foxholes with the mud, rain and freezing cold. I tried to explain to them what it was like up front, but my tooth hurt so badly that it was difficult for the medic to understand me. They looked at me as if I came from the planet Mars. They shook their heads and said "Oh my God" with looks of amazement.

One of the medics asked me if I would like something to eat. I asked for some soup. I hadn't eaten since the night before. He came back with a can of soup and some fresh bread. When the medic opened the can of soup with a key, it automatically started to heat up by itself. I asked how it worked and they said the British made it. I wished we had that type of soup up front. After I ate the soup, ever so carefully because of the pain in my tooth, the medic gave me some new clothes and told me to go into the bath house. Someone took me there. I undressed and enjoyed taking a shower. It felt so good, warm water and soap! It sure was better than washing out of my helmet.

I put on the new clothes and went to the dentist, who was waiting for me. From his office he had seen me come in earlier, all covered in mud. He looked me over and asked me the same questions that the GIs had asked me in the hallway. I am sure the GIs told him what I said to them about combat. He looked into my mouth. I pointed out the bad tooth and he told me that it was infected. He gave me some pills to take, but said it was too swollen to pull the tooth out that day and he would have another look at it in the morning. It was badly infected.

The men gave me a canvas cot to sleep on. When I fell asleep, it was like heaven, it was so peaceful. The medic gave me a few aspirin before I went to sleep. I slept so well. There were no enemy guns to worry about. I felt like I was back home in the States. The next morning, I got up and washed and cleaned myself. Then, I had a canteen cup of coffee.

After drinking down the coffee, I went in to see the dentist. I sat in the chair and he gave me a few injections in my gums. After my mouth was numb, he pulled the infected tooth. Then he put a sterile pad on the place where there had once been a tooth. He explained he would make me a partial plate and then said, "Lots of luck, Lauria. I hope you make it through this war."

All of the boys who looked after me while I was there also wished me all the luck in the world. I thanked them and said I could use it. I got back on the truck for the ride back to the front lines. They waved to me as we pulled away. The ride back was quiet and calm at first but it wasn't peaceful for long. It was only a short time before I heard the sound of artillery and machine-gun fire and I started to get afraid again. The day I spent in the rear was like a dream, and it had come to an end.

At least I got a good night's sleep and some clean clothes to wear. It only took a couple of days before the new clothes looked as bad as the ones that I left back at the aid station. The partial plate the dentist had made arrived about a month later and I learned to wear it.

I arrived at my company command post and some of the troops greeted me. The outfit had advanced a few miles while I was gone. They did a good job without me, ha, ha. I was happy to be back with my outfit for one reason. Many of the men that got wounded would never get back to their outfit. Sometimes the army screwed up and sent them somewhere else. After being in my outfit for two years, I had grown attached. This was my home away from home.

Hedgerows

We fought for the control of hills. One day they would be ours, and the next day the Germans pushed us off. I remember my sergeant told me and another crewmember to run a line to the guns that were about a half mile in back of the command post. As we went along the narrow road with hedges on both sides, mortar shells dropped a couple of feet from us. Pop Rocky, my partner, and I dropped to the ground and waited for the shells to stop. Rocky was about 40 years old and we called him Pop Rocky or Dad. Then we continued laying the line that led to the cannons in nearby fields. Another round of about ten more mortars landed several yards from us.

After the shelling stopped, I completed my work and walked over to Lt. Roggenstein. He was in a ditch along a hedge. He called to me, "Lauria, get your ass in this ditch. Are you crazy?" I said, "No sir, just doing my job trying to get this line over to the guns and back to you." He said, "I know that, but wait until the shelling stops." I thought to myself that waiting would take a few hours. The lieutenant wasn't too gung ho. I don't blame him. He was a good officer. After ten minutes of waiting, Pop Rocky and I headed back to our observation post on the front lines.

As we went up to the front line on the same road where we laid out the lines earlier, Sergeant Turner told us we had to put an overhead line to go from one side of the road to the other. So I started to climb a tree to hang a wire in it. Then I climbed a tree on the opposite side to hang a wire on that one too. I was pretty high up in the tree because the wire had to be high enough that tanks and trucks could clear it. While I was in the tree, about ten German mortar shells showered down around me. I was terrified. I slid down the tree as fast as I could, but it seemed like an eternity. I said as many Hail Marys as I could muster and I dove behind a nearby high hedge made of stone and dirt. One of the shells landed on the other side of the road exactly where I had climbed the other tree.

The noise from the explosions made my ears ring. I burrowed through the hedge trying to find cover from the surrounding explosions. Fear made me as strong as a bull. I landed on two men from my company. They thought that I was part of the debris from the explosions. The looked at me and said, "Lauria, where did you come from?" After I collected myself, I told them that I had been laying line on the other side of the hedgerow. They both looked at me and shook their heads and told me that they had thought that I was a tank coming through the hedgerow. They couldn't believe that I made such a large hole through the hedgerow. I still can't believe that I did. I guess when a rugged little guy is scared, anything is possible. After the shelling stopped, I called out for old man Pop Rocky to see if he was safe and alive. Rocky called back and said he was okay so I picked myself up and went back out the hole I had made. Pop Rocky and I started to head forward again to the O.P. without any further interruptions.

Saint-Lô

Fighting in Normandy was one hell of a nut to crack. The going was slow. Some days we were lucky if we could gain a few hundred yards. The hedgerows presented a difficult problem. If you tried to climb over them, the Krauts would spot you and shoot you down like a clay pigeon. The hedgerows also had booby traps in them such as trip cords. We had a high number of casualties in Normandy's hedgerow country and the going was slow.

After a month of this type of warfare, our advance was almost at a standstill. I believe the big brass had plans to do something about it because casualties were so high. We were dragging our feet and things were not getting better. Along the front, three or four divisions were told to move back about two miles or so and hold fast. I had no idea what was going to take place, but we were only about a mile from Saint-Lô. The city was a hub for all routes going in every direction. We aimed to break through and spread out in every direction. The road would be great for tanks. At this time, we did not know what the outcome would be. We spent the night in crowded conditions and we were extremely cold. One blanket was not enough to keep warm.

That night we slept in an open field in foxholes. We were up at the crack of dawn. Troops were crowded together like sardines. There were many different troops mixed in with our unit, that's how crowded it was. I couldn't believe how many doughboys were in these fields. I remember eating some cereal which I put into my canteen cup with water—there was no such thing as fresh milk—but I did enjoy it. It came in a small box the size of the K-ration, but I never came across this cereal again.

Around nine in the morning, fighter planes flew over our heads. They flew about as high as tree level or the height of the hedgerow. You could almost touch them with your hands. The noise from the engines was deafening. They scared us half to death. The planes surprised us and before we knew it they were flying directly over our heads at three hundred and fifty miles an hour. The fighter planes tried to knock the enemy ground troops out. This activity went on for several hours. Then there was a short lull for a few minutes.

We heard a bunch of B-17 and B-24 bombers flying around twenty-five thousand feet. The engine noise made it difficult for us to comprehend anything. Squadron after squadron went overhead. A squadron consisted of about 36 planes. The first bomber spotted the target and released a smoke bomb along with other bombs. All planes in the squadron dropped their bombs where the lead plane marked the target with smoke.

The Germans came out of their holes and began to fire their anti-aircraft guns along with 88mm guns. Many of the planes were hit. When they hit the ground, the earth rumbled as if there was an earthquake. We prayed that they hit their target because if they missed there was a good chance that we would get hit.

I looked up at the sky and counted the number of American planes that were shot down. In the four hours it took to complete the raid, I counted 98 American B-17s that went down. The bombing raid made the pit of my stomach sink because I witnessed so many pilots' deaths. I didn't see any parachutes open when the bombers went down. The planes would tumble and roll about, somersaulting. Sometimes they went down in flames in a dive and exploded. Fire and smoke enveloped most of the planes on their way down.

After 56 years, I can still picture the American planes plummeting down. Each plane had nine or ten men in it. This means that over nine hundred airmen lost their lives in a single raid, and this was only one small event that I witnessed. After the bombing ceased, the remaining planes circled the city of Saint-Lô and mopped up any enemy activity they could spot on the ground for another hour. Many American troops lost their lives because of our own bombers. Some bombs landed short of their targets and hit American GIs. One of our generals was killed this way. He was too close to the bombing activity because he probably wanted to get a better view.

After the air raid, we received the order to mount up and head for the city of Saint-Lô, which was nothing but rubble. Army engineers cleared the way for tanks and ground troops to reach the main roads. French civilians helped clear the way and as the army dashed through,

The planes bombing St. Lô.

they yelled to us, "Thank you America! Thank you." We began the process of moving across France. After going through Saint-Lô, plans to form a Third Army came together. At first, I was part of the First Army, but after the breakthrough at Saint-Lô I became part of the famous Third Army.

I made some sketches of what I saw at Saint-Lô. The sketch only shows a few troops but the field was really jammed with troops like sardines. I didn't have the patience to sketch so many men. Saint-Lô was an important area because it was a hub for major routes going in all directions. Once we broke through at Saint-Lô, we could branch out in all directions and then spearhead to the south of France. I guess the high brass had something in mind for us little guys. I just followed orders.

Our outfit continued the process of covering ground in France. The weather got much warmer and the troops were sweating constantly. We all hurried along the narrow French roads as if we wanted to be the first to reach the enemy. The Fifth Division tanks rode alongside of the 11th Infantry Regiment. Our unit took the narrow roads. We traveled south at about 30 or 35 miles an hour.

This war was great for tanks, but the tanks made so much dust that

46 Running Wire at the Front Lines

Illustration of the Fifth Infantry Division spearheading south through France. Sometimes we would go anywhere from 25 to 50 miles a day, taking town after town. There were miles and miles of dead Germans' horses and destroyed jeeps and tanks from the P-47s and P-51s.

we had to wear dust goggles. The tanks alongside us were a comforting sight and they gave us a welcomed sense of security. The dust was so bad that you could see it for miles around. Thank God the German air force didn't cause us any difficulty. The American air force appeared every now and again. When our planes arrived, it made our unit feel good to know that they were around to protect us.

Our unit stuck out like a sore thumb, so it was nice to have extra protection. There were times when we went too far ahead and our air force would come down on us to attack. When they saw the big star on the vehicles or tanks or on the panel on top of the vehicles, which was a code, they would know we were their boys. Panel colors changed whenever the regiment and air force gave the orders.

The GIs traveled in the back of trucks or hung onto the sides of tanks. The poor riflemen would be on top of the tanks, hanging there for hours. I was lucky because my crew had a small wire truck and we were sure of a ride. We also put some riflemen in with us.

Six 109 Messerschmitts headed straight at our column of vehicles.

Every once in awhile, our unit came to a halt. The doughboys got off and took a break. Our eyes, throats and mouths were full of dust. We wrapped handkerchiefs or cloths around our faces for protection. The dust even clung to our skin, making us look like mummies. It was all over our clothes and face, as if we had taken a mud bath. We only stopped for darkness or enemy activity in the distance. The trucks drove on the roads and the tanks rode alongside in the fields, churning up the dust as they went. It was August fifth or sixth, and it was very hot and dry, creating a lot of dust. When we did stop for a break, our unit was always exhausted. We dug foxholes for the night's stay and if we were lucky, we wouldn't have to perform guard duty.

At this point in time, the Germans were disorganized due to the heavy bombing at Saint-Lô. The German forces there had to flee for dear life and try hard to form a new line of defense. We met some pockets of resistance on our journey south.

Things looked good and all units were on full alert. The tanks and jeeps had 50-, 20- and 30- caliber machine guns mounted on their vehicles. They kept a sharp eye up on the sky. There were some low clouds lingering around one day as we were heading south when suddenly all hell broke loose and guns opened up, shooting into the sky. Six German 109 Messerschmitts were headed straight at our column of vehicles. Somebody yelled out, "Air raid!"

Most of the company ran for cover and vehicles skidded to a halt. My wirer crew jumped off the vehicle and ran. On the left of the road there was a ditch with a large corrugated pipe, which was covered with a walkway made of wood that served as a path to a home on a small hill. As the planes dove down at us, I tried to dive down into this pipe. Luckily, it was wide enough for me to squeeze into it, but two of my buddies were already inside. I pushed and pushed, trying to make room for myself. One of them yelled at me to quit pushing them, but I kept trying to get into the drainpipe.

The six enemy planes dove straight towards us. Some men hid under trucks while others ran into the open fields. The planes used 50-caliber guns and their bullets just missed my half-exposed body. When a bullet hit the ground, a rock shot up and hit me on the behind, but I was unharmed.

Some of our men fired back at the planes but didn't hit anyone. The enemy 109 planes took off for home. This certainly was a close call for our unit. There were no casualties. After everything cleared, we mounted up and continued on in the same direction. After this air attack, we didn't encounter many more German airplanes, because the U.S. air force did a wonderful job of maintaining air supremacy. I believe that if the Germans could have matched U.S. airpower, the Allies would still be waiting back in England and Ireland.

IV

Angers

Combat was a life of stress and hardship. We fought not only the enemy, but also the weather. We endured heavy rain, snow, mud, and most of all terribly cold nights. It is hard to understand the feeling of combat unless you yourself have taken part in it.

After the breakthrough at Saint-Lô, we advanced south through France, fighting our way from town and making river crossings. The division arrived on the outskirts of a small town called Angers with a population of about eighty thousand. Our unit was held up there by the enemy's small arms and machine guns.

Most of our doughboys rode on tanks and trucks because of the speed at which we advanced. Hearing gunfire, the doughboys jumped off the vehicles to take cover in the ditches and under trees. A few seconds passed and our riflemen began to fire on the enemy. There were about 20 minutes of exchanging gunfire before the Krauts fled.

Our unit had advanced about another mile forward when the captain gave orders to come to a halt because of darkness. Arrangements were made to set up sentries to guard our position against night attacks. That night some small-arms fire and shells landed on our position but did very little damage.

The next day our unit received orders to mount up and move out. We advanced about a half mile and our captain ordered us into a field surrounded by hedgerow, which made matters worse. Little did we know, the Krauts were dug in about 40 feet away in the next field. They were set up to take us all down.

Our Cannon Company got into position while I prepared to string my wire reel from gun to gun. Just as I started to get off the truck, the Germans opened up with burp and small-arms fire. The shells went across the hedgerow while I was standing upright on the truck. I could have been cut in half by the burp gun fire. The bullets crossed past me and

Running Wire at the Front Lines

Two Germans had a burp gun in a second-story window which held us at bay. One of our foot soldiers blasted the window with a submachine gun and another fired his rifle grenade at the window, putting those Germans out of action.

luckily I was able to jump into a ditch alongside the hedgerow. Our riflemen returned fire for about two minutes before the Krauts began to flee. The "all clear" was called out and I continued wiring each gun.

In the meantime, the driver of the truck, Dutkiewitze, was in a ditch and shaking uncontrollably. He was so shaken that he refused to get out of the ditch and drive us down the road to lay the telephone line. Sergeant Turner arrived at the ditch and told him to get out, but Dutkiewitze refused. The sergeant asked for a volunteer to drive the truck. Pfc. Turner accepted the job, so I got back on the truck and began to lay the telephone line.

About a half mile down the road, our unit halted. The doughboys

Opposite Top: Me standing upright on the truck as the bullets passed by with Dutkiewitze in the ditch nearby. *Bottom:* Outside Angers, waiting for the tank battle to end so the riflemen could move on to kill the German riflemen.

waited alongside the road in irrigation ditches. A tank battle took place for some time and there were heavy losses on both sides. We lost two tanks to land mines, and one tank was lost in an anti–tank ditch. C Company and G Company sweated out most of the day while the rifle company regained control by nightfall.

I returned to the command post. I had a C-ration which was a can of corned beef hash and a cup of coffee. The hash tasted more like cold dog food. Then I got some sleep, but not for long. Our telephone line must have broken somewhere near the tank fight. My partner, Chester Pilip, and I had to set out to repair the lines. We stopped every 15 or 20 feet, tapped into the wire and checked it. Then we followed the line to where the tank fight was taking place. Crawling on our hands and knees and checking the lines, we had to hold onto the wires because it was so dark we couldn't see in front of our faces.

My buddy Pilip called to me, "Lauria, I feel something like a body here." A few seconds later I also felt a body. The hair on my head stood up straight. I said to Pilip, "Let's get out of here." We didn't know if they were Americans or Boche. We called to the command post and told them it was impossible to find the break in the line because of the condition of the minefield. We were ordered back to the command post.

It wasn't much fun being out there in the dark alone. You get so many things going on inside your mind. You wonder if the enemy is behind you ready to shoot or stab you. When we were out in the dark, we went through so much stress because we wondered who might shoot us. Once a sergeant named Hill was shot by one of our own men in the dark, so we had to be very careful out there at all times. After we got back to our post at four in the morning, we managed to get about one hour of sleep before they told us to go out there again and try to repair the line for the morning attack. So my partner and I set out again, taking with us an extra wire and a D.R.-4 reel with almost a mile of line on it. We spliced the line just before the minefield and continued to where the attack would take place.

We reeled the line to a peach orchard and waited for dawn because that was about the time our troops would start to attack. Pilip and I had had almost nothing to eat in the past 24 hours but we couldn't eat the peaches because they were not fully ripe. I tried to eat one anyway. It was so bitter that I spit it out. The captain laughed at me and said, "Lauria hungry?" I said, "Yes, sir." He answered, "So are we." The captain did offer us a cup of coffee and we gladly took it. It sure tasted good. As we waited for daybreak, Captain Smith, Len Tredanari, Pilip, Sergeant White and I huddled together and talked about the attack that was going to take place.

They finally sent word for us to attack. On August 9, the First and Second Battalions of the 11th Infantry jumped off. The company commander yelled to everyone to get a move on. It wasn't easy to just get up and go. We had to climb a seven-foot stone wall. Some of the riflemen jumped off the top of the stone wall and dropped to the ground, as I did. On the other side of the wall was a line of houses. We headed to the right to advance south and came across sporadic gunfire. At this point, casualties were light and we advanced about 100 yards down the road and made a left, heading east. At that moment, our doughboys were engaged in a heavy artillery battle. Foot soldiers landed all around us and gunfire came from every which way.

Our doughboys fought from house to house, which we called mousehole fighting. The Boche fired from windows and doors. The Germans had a burp gun in a second-story window which held us at bay. One of our foot soldiers blasted the window with a submachine gun and another foot soldier fired his rifle grenade at the window, putting those Germans out of action.

After that, our men went around the corner down the next street, towards a small bridge that had been blown apart by the Boche deliberately to hold back our advance. The high explosives did their job destroying it. Our doughboys carefully walked over the obstacles and craters. Pilip and I followed the riflemen across. We laid our telephone line along the outside of the bridge away from the center so our tanks would not destroy it as they went past. Pilip and I stayed close to the doughboys. As we crossed over the bridge, a Frenchman yelled something to me. I couldn't understand him because I don't speak the language, with the exception of a couple of small words. I started to walk again and had gotten a little further when I realized what the Frenchman was trying to say. There in front of me was a mine. I just missed it by inches. I turned to the Frenchman and yelled, "Merci beaucoup!"

I called for the army engineer and he dismantled the mine. We started to advance again and the foot soldiers picked their way from house to house, flushing out the Germans. Pilip and I stayed alongside the riflemen as we laid down our line.

As we went along, the fighting got more intense. It seemed like a hundred Fourth of Julys. Mortar shells roared and exploded over and in front of us while we ducked in and out of hallways. A Frenchman was running across the street when a shell exploded a few yards away. The shrapnel from the shell cut his leg off about six inches above the knee. His screaming was unbearable. A few Frenchmen ran out to help him. One man ran out with a push cart. They picked him up, put him into the car and wheeled him to the aid station.

Our medics tried everything in their power to save the Frenchman's life. He passed away in five minutes from the loss of blood. Our aid station was set up in a vacant store and was really for minor wounds. If the wounded could be evacuated, they were taken to the army field hospital somewhere in the rear. With blood splattered all around me, I moved to a better location.

The First and Second Battalion kept up the advance as mortar shells came in from everywhere. At last, we arrived at the main intersection, which was about 150 feet wide. To the left of the intersection was a dead end about 15 feet back. It was a good advantage point for the Germans. When they opened fire with 20- or 40-millimeter guns, the shells that missed the troops would hit the wall, ricochet and explode, increasing the chances of hitting the doughboys.

The German 20s and 40s were only about 100 yards down the road and were concealed very well. It was hard to hit them with anything except cannon fire. So we were pinned down on the side of the road, waiting for orders from our captain to cross over the intersection. The captain had already crossed over about an hour before with Len Tredanari and Sergeant White. The radio wasn't working due to the interference from the buildings. Our wire was so important because without good sight of the enemy the cannons were useless. The captain gave orders to Sergeant White to head back to Pilip and me and have us cross over. The sergeant gave us the orders, but I must admit I was nervous and afraid.

While I looked at the intersection and waited for orders to cross, it seemed longer and longer. The more I waited, the more I sweated. Whenever a soldier would peek around the corner of the house, the Jerries would open fire on us. The shells would hit the three sides of the dead end and shrapnel would fly all around us. Some GIs got across only by inches. A sergeant from Company G ordered a few of his men across but they froze in their tracks. The sergeant picked his rifle up by the barrel and swung it at the riflemen. Six of them decided to take a chance and dash across. It was a good thing they were all lined up shoulder to shoulder. The shell

The intersection in Angers where we waited to cross over with out telephone line. The wounded GI yelled out at me not to attempt the crossing.

passed their backs by inches. I saw the shell go by and made the sign of the cross the second they made it safely to the other side.

While we waited for orders to cross, I wanted a drink of water because I was so nervous I couldn't swallow. A Frenchman was standing in the doorway and I tried to speak to him about getting some water. I motioned to him with my hand cupped to my mouth. He said, "oui, oui." A few minutes later he was back with a bottle and I took it and started to drink. It happened to be cider, hard cider, and very warm. I swallowed a mouthful and spit out some. I was in no mood to be drunk.

While I waited for orders to cross the road, another Frenchman came up to me and said that a German sniper was somewhere in his backyard in a tree. So I followed him through a hall-like tunnel to the back. As soon as I got to the opening, there was the tree, full of leaves. I couldn't see anything. I moved to my left with my rifle and there was a shed about 15 feet away. All the Boche could see was my waist up. I looked again and about five shots rang out. The Kraut fired at me, but missed. The shots went on both sides of my body. I can't understand how he missed me. I believe that the all-mighty God bent the shell to miss me. It was a miracle.

I was so angry with the Frenchman. I wanted to hit him with my rifle. I was yelling at him when the sergeant called my name, "Lauria, where are you?" I stopped yelling at the Frenchman and then answered my sergeant, "I'm here." Sergeant White said, "What the hell are you doing back there?" I told him the story about the Kraut in the tree. He started to give me hell for being back there looking for the sniper. He told me I had more important things to do. I guess he was right. It was a rifleman's job to kill that S.O.B. I was so angry at myself for doing such a stupid thing. I liked Sergeant White. I had a lot of respect for him.

While I got ready to cross over the intersection, a rifleman approached me. He looked very sad and angered. He told me he had just gotten a letter from home saying his brother had been killed in action over in Italy. Being so emotional and angry, he said he killed 20 Germans in this day's attack. I put my arm around his shoulders and tried to console him. At that very moment, a German sniper shot at us from a house on the other side of the intersection. The rifleman said to me, "I'll get that son of a bitch!" and dashed across the road, not thinking of the danger that lay ahead. He ran into the house where the shots were coming from. A few minutes went by and I heard two shots ring out; seconds later the GI crawled out into the center of the street yelling for a medic to come and help him. Two medics ran to him and saw he was shot in both legs. They gave him a shot of morphine to help kill the pain.

In the meantime, Pilip and I got ready to attempt crossing the intersection. The wounded GI yelled out to us not to attempt the crossing. As hurt as he was, the GI was concerned for us and the dangers ahead. Sergeant White was at the corner building. Pilip and I laid out about 150 feet of wire behind us so that when we ran across it wouldn't jam in our reels. Sergeant White called to us, "Are you guys ready?" We nodded our heads yes and waited for the signal. We kept our eyes on the sergeant's arm because the minute he dropped it we would run across.

As we waited for the signal, my body shook like a leaf. Finally, the sergeant dropped his arm. We began to run with me on the right of the reel and Pilip on the left, holding the shaft. Being so nervous, my legs went twice as fast as Pilip's. He called out for me to slow down. In the corner of my eye, I could see his feet were up in the air like a kite in the wind, with me at 5' 4" and Pilip at 6' 2". They say when you are nervous you can do twice the amount you are usually capable of doing.

About halfway across, the Krauts opened fire on us. The 20- and 40-millimeter shells were so close to us that we could feel the heat on our backs. We made it safely to the other side. I was sweating like a pig. My clothing was soaking wet. I knew I had pissed my pants from all the excitement and I am not ashamed to admit it. It had happened before on another mission.

As soon as I got my head together, I stopped to talk to the GI who was wounded. The medics were putting him on a stretcher. The GI told Pilip and me that we were crazy to attempt the crossing. In return, I wished him Godspeed and good luck and we said our goodbyes. In a crazy sort of way, I was happy for him because it was like a million-dollar wound and maybe he would be sent home.

As we advanced further, the doughboys flushed the Krauts out of the houses. Some of them came out with white rags over their heads, surrendering. I moved along with Pilip and told the Krauts to keep going down the street. Sergeant White told us to follow him through a three-story house with an attic. The captain told us to bring the wire with us. Captain Smith and Tredanari were waiting for us in the attic. The captain said to us, "Glad to see you boys made it, good work!" Pilip hooked the line to our field telephone and called the command post.

The observation from the attic window was perfect. We could see the Krauts clearly as far as 100 yards away with their guns pointed at the road block. The captain requested the fire mission to shoot one round of smoke to pinpoint the shell explosion. Luck was with us because the shell landed within 50 yards of our target. The captain requested that the Cannon Company shoot all six guns at will. For the next two minutes, the guns fired a good number of shells that landed on the enemy, putting them out of action. This made it possible for our troops to cross over to the other side of the intersection. After a few hours of being flushed out of the houses, the Germans all surrendered willingly.

Krauts were giving up, coming out of the buildings with their hands on their helmets and waving white rags. Pilip and I kept yelling "You S.O.B.!" at the Germans. I wanted to kick them in the ass as they went down the street. I guess the Krauts were happy to be taken by us because the war for them was over.

Sergeant White told Pilip and me to follow him through some build-

I stopped in this large church and prayed to God for all he had spared and thanked him for putting an end to three days of stress and combat in Angers.

ings. A few French ladies in their fifties saw me and thought I looked so young, more like a baby. One made the sign of the cross and kissed me on both cheeks. We both had tears in our eyes. I continued a few more steps and the French lady said to her friend, he is only a "bébé," which I understood. She then put her arm up in the air and said "Viva l'America!" She also said in English, "God bless you all."

The captain told us to head back to the command post. On our way back down the same street we came from, I stopped in a large church and prayed to God for all he had spared and thanked him for putting an end to three days of stress and combat.

After we captured Angers, many houses fell into our hands, including all kinds of equipment and a warehouse full of alcohol, wine, and champagne. You name it, it was there. I am sure most of the boys in the outfit got drunk for the next two days. I remember everyone in our regiment had a jacket made of rabbit fur that the Germans had left behind. The Germans withdrew in a hurry and left all their equipment for our boys to enjoy. Our wire truck had about five cases of liquor in it. I didn't drink, but I did enjoy seeing all my buddies drink wine and champagne.

After I left the church, I headed for the truck, which was about a half mile away. As we went down the street, the French people ran over to us GIs and called out in French and English, "God bless you," and "Viva l'America." It was a good feeling to have them give us flowers and bottles of wine or cider. This took some of the stress off me. Most of us looked like zombies in the daytime. We were wet from head to foot from heat and fright.

When I arrived at the wire truck, I put my equipment down and looked up and saw Dutkiewitze sitting there, drunk as a pig. He wanted to drive Pilip and I back to the guns, which may have been about a half mile down the road. Sergeant Turner was also there with Pfc. Turner. I looked at my sergeant and said, "He is not going to drive this truck. If so, I will walk back." Sergeant Turner then told Pfc. Turner to drive us back to C.P. After that, Dutkiewitze lost his job as a driver and was put into the cannon platoon.

I was so angry with him. Not because he was afraid to get out of the ditch when the Germans were firing at us in the hedgerows a day or two back. I was pissed with him from way back in Normandy. The first night in combat he and I had shared a foxhole that was shaped like a V. We had shell cases filled with dirt on top of our foxhole, which made it a strong, safe place to be. Just about midnight we got into our foxhole and all hell broke loose. Shells rained in on our position hour after hour. The ground shook. It seemed like shells exploded only a few feet away. We both held onto each other, praying. Round after round dropped on us. It went on all night long. We both prayed and cried out loud.

When morning arrived, we were happy to be alive. My pants were wet with piss. I got some coffee and a K-ration. After having my coffee, my mail sergeant asked me if I would like to go with him for the mail back in the rear. I said yes, so Corporal Bell, who was our jeep driver and about 4'1", told the sergeant and I to jump in the jeep and said, "Let's go." So we took off. Little did I know, it would be a dangerous ride because the road was exposed to enemy artillery and mortar fire.

As we drove for the mail, Jerries fired at our jeep and a tank that was exposed when we went around a turn. A few men went back with us to our C.P. We went the same way we came and Corporal Bell waited for

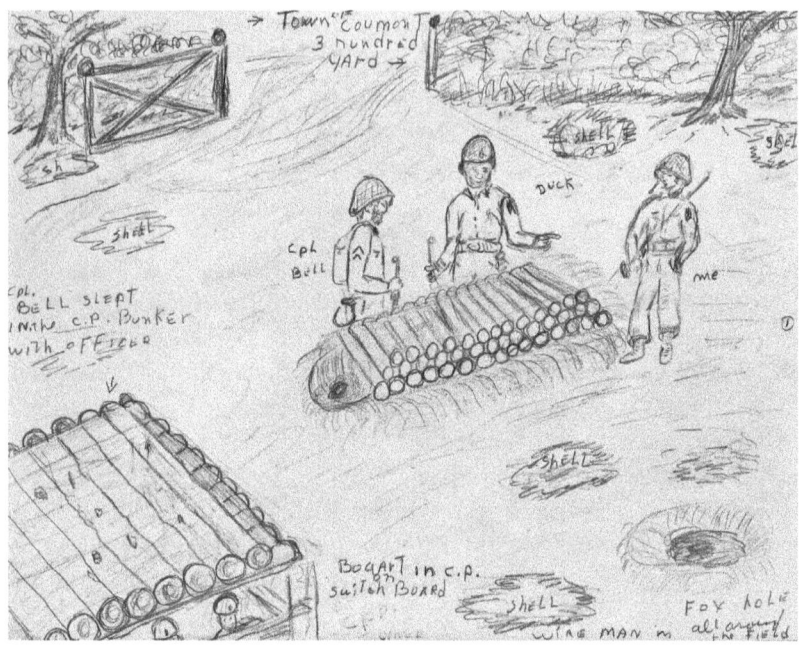

In Caumont, when I overheard Dutkiewitze tell Corporal Bell, "Lauria wants to act all brave because he took a ride with you." He then told him how I was acting in the foxhole all night, shaking, praying, pissing and crapping my pants. The corporal answered Dutkiewitze, "Aren't you ashamed of yourself? You want to compare yourself to that kid? He is only 19 years old and you are over 30."

the M.P. to give us the go-ahead. At the dangerous point, he stepped on the gas, so we made it safely back to the others.

Getting back to the C.P., we had mail call. Mail was given out. I received a few letters from home. While reading them over, I overheard Dutkiewitze tell Corporal Bell, "Lauria wants to act all brave because he took a ride with you." He then told him how I was acting in the foxhole all night, shaking, praying, pissing and crapping my pants. The corporal answered Dutkiewitze, "Aren't you ashamed of yourself? You want to compare yourself to that kid? He is only 19 years old and you are over 30." When the corporal finished telling Dutkiewitze this, he turned around and walked away.

I felt so bad and ashamed of myself. Bell said, "Don't feel bad, Lauria. We are all in the same boat. I was also afraid." I never told Bell what we both went through that night until a month later. I was so angry with

Dutkiewitze, not for getting drunk and leaving his post, not for being scared, but for telling Bell about the foxhole deal. I know you can't control yourself in combat when you're so green. It plays tricks on you.

After this incident with Dutkiewitze, I kept my distance. I never really got friendly with him again. Two days later we were told to mount up and move out. Our orders were to move northeast towards Chartres.

V
Chartres

After the fighting in Normandy hedgerow country and breaking out at Saint-Lô, the Fifth Infantry Division moved toward the south of France, fighting in small towns and cities. We advanced with two other regiments and took a lot of ground. Lots of little towns fell into the division's hands one by one. As the outfit went deeper and deeper into France, our lines of communication were getting longer. Most of the men in the company were in constant fear that the enemy would cut our unit off from the rear as we fought and pushed deeper into France. We stuck out like a sore thumb.

Since I was a radio wire man, I was given maps of the French terrain, so I knew that our next objective was going to be the city of Chartres. The enemy was going to fight hard to hold this city as long as possible. The 11th Infantry arrived on the outskirts of Chartres around the 19th of August.

Most of the doughboys rode on our trucks and tanks or any other vehicle they could get on. We sometimes advanced 20 or more miles a day. One day when it started to get dark the convoy came to a halt. Doughboys jumped off the vehicles and started to go forward. The Cannon Company crew did likewise and put their cannons in position in the direction of the enemy. In the meantime, riflemen walked in scaled groups into enemy lines. The enemy fired at our riflemen, but our doughboys kept walking forward, firing back with rifles and Browning automatic rifles or B.A.R. This was what we called marching fire.

I had to get on my hands and knees to string my wire to the guns. Bullets rang over my head. Some crew members were digging in, but I didn't have time. The guns came first.

I remember it was just about sunset when the riflemen walked past our gun position. Enemy shots went overhead. The sun at this time was at its lowest ebb. The GIs kept walking ahead and continued their march-

It was just about sunset when the riflemen walked past our gun position. Enemy shots went overhead. The sun at this time was at its lowest. The GIs kept walking ahead and continued their marching fire.

ing fire. It had gotten much darker. I could still see our boys' silhouettes fading away in the distance towards enemy lines. There was just enough light to see our doughboys passing the gun position. They walked in marching fire closer to the city.

In the meantime, some of us from the wirer section went forward to meet up with the forward observation post, where Captain Harry Smith, Sergeant White and Pfc. Tredanari were on the radio. The captain proved to be a hell of a great leader in our company. Captain Smith and Sergeant White looked for a good location to set up an O.P. When we were about 400 yards closer to Chartres, we held up in a barn. It was too dark to go any further, so the barn was the observation post for the night. Lots of doughboys stayed inside the barn and others stayed outside and around it. The weather was hot, so it made no difference.

After I hooked up my line inside the barn, I was told to go back to the guns. Campbell stayed in the barn with the captain, Sergeant White and Tredanari. The riflemen took turns guarding our newly won position.

The crew inside the barn did likewise, taking turns guarding the observation post.

Tredanari set up his radio and made contact with me back at the command post. I received his call. His code name was Dusty Black 6 and I was Dusty Black 2. I repeated, "Dusty Black 2 to Dusty Black 6, I hear you loud and clear." Tredanari told me to stand by for a fire mission. I alerted the gun crew and then told the number one gun to stand by for a fire mission. All we could do now was get a fix on our position and pinpoint a target.

The number one gun got his coordinates from me and I got them from the executive officer. I repeated it to the number one gun. Also, I received the bass deflection, elevation and charge to use. It could be a high explosive or a smoke shell. A good five to ten shells may be used to pinpoint a target. It was difficult to locate the explosion in the dark. When we located the target, we ceased fire and waited until morning since the enemy didn't retaliate.

In the meantime, back at the command post, the gun crew sergeant was making arrangements for one man from each gun emplacement to guard his gun. Each man got a turn on guard duty. I did my two-hour shift of guard duty that night and then tapped the next man to take my place. There was some light machine-gun and rifle fire during the night but most of our company slept. At night, I heard Germans speaking but I didn't wake up to see their feet going by my foxhole. Being so exhausted, I didn't wake up. I heard machine guns going off, but feeling secure and knowing we had our guards on duty, I just slept away. There was some yelling and some spurts of machine-gun fire just a hundred yards away but I didn't pay it any mind.

The last guard woke me up around six bells. He told me that during the night at about two A.M. a German patrol had walked into our gun area without making a sound. Most of our boys were asleep. The guard on duty didn't recognize that the patrol was German. The patrol had just walked past our gun position. We learned about this the next morning. The German patrol had gone across the field into a barn about two hundred yards from the gun position. When they got to the barn and opened the barn door, they were speaking in German. When the guard on duty called out, "Who goes there?" the Krauts were in shock. The GI and a few other doughboys opened up with rifle fire and killed all six men.

The word got to Lt. Roggenstein. He wanted to know who was on guard duty at that hour. The man that was on duty at that time said he had seen men walking near Lauria's foxhole but thought it was our company men. I was lucky everything turned out okay. Most of the time enemy patrols didn't go looking for trouble. They just looked for gun placements and whatever information they could go back with.

The captain told the last man on guard duty to wake everyone up at daybreak. Back at the observation post, Sergeant White was the first one awakened by the guard. He rubbed his eyes and moments later walked over to a hole in the barn wall where some of the boards were broken. He peered through this hole and saw nothing. The morning mist was too heavy, so he waited for the fog to clear. Sergeant waited and paced the barn floor for about ten minutes before looking out the hole in the barn again. The fog was clearing up.

He waited about ten more minutes and took another look. He said, "Oh my God!" and called Captain Smith to take a look. When Sergeant White looked through the hole in the barn, he saw forty 20- and 88-millimeter guns on a 150-yard front by a wooded area. The weapons were located in front of a church. The Germans had Red Cross flags draped over each gun. Captain Smith immediately told Tredanari to get the Cannon Company on the phone.

I was back at the command post by the guns, which were only about 200 yards away, and got the call to stand by for a fire mission. I yelled out at the top of my lungs to all the guns, "Fire mission!" Within a second, most of the gun crew was at its cannon position. We woke up the executive officer, and I gave him the fire mission's location. He had a map and slide ruler. He then gave me the elevation S.I.E. and charge to use. I gave this information to the number one gun, who used one round of smoke. I received orders from the executive officer to have the guns fire. I repeated it to number one. Then I called O.P. and told them that one round of smoke was on its way.

The smoke shell made it easy for the observation crew to spot the location much sooner. Luck was on our side: the shell landed dead center of the target. The Krauts saw the smoke and ran out of their foxholes and the church, trying to get to their guns. They knew there was an observation post in the barn. I received a second call from Tredanari in the O.P. telling me to hurry up and have all six guns adjust with high explosives; the number one gun was on target. All guns were ready and given orders to fire at will. Some 60 to 70 rounds were fired.

After a few minutes, I received orders to cease fire. Seconds later, Tredanari called back and requested a fire mission just a few degrees to the left. Again, the captain wanted the guns to fire at will. After about 50 rounds, a call came back to me to cease fire. I told all guns to cease fire and everyone waited for the results.

When the smoke cleared, Germans were splattered all over the area. Most of their garrison lay dead. Other Krauts, who were not wounded, ran into the nearby woods on both sides of this 150-yard front. They were also killed because we used quick-fire shells that burst into the treetops. They have the same effect as a time fuse. When the shells burst in the treetops shrapnel scattered downward into their foxholes, killing and wounding most of the garrison. The Krauts were crying for help and calling for their mothers as they lay dying with no arms or legs. The Cannon Company fire mission had killed about 250 Germans in a very short time. It only took about 15 minutes to do so much damage.

The barrage from our guns also blew up an enemy ammunition dump. K and F Rifle Company captured the entire German garrison of 800 men mostly because the Cannon Company had done such a great job getting the first shot dead center on the enemy guns. The 11th Infantry suffered four riflemen killed, one officer and twenty enlisted men wounded. It was a miracle. The outfit's training in Ireland paid off. As we went ahead deeper into France, the Fifth did many heroic deeds. We always accomplished our missions. We had pride in our outfit and performed well most of the time. We made 26 river crossings, and I don't mean over a bridge. We had to go across each river with assault boats. I often wonder how I ever survived.

Chartres had the most beautiful cathedral in the world. It was untouched by war but some other houses of worship in Chartres were damaged and damaged badly. The Cannon Company had to knock out a church steeple for the simple reason that it was used by the Germans for observation. As the saying goes, "All is fair in love and war." It works both ways.

In the town of Chartres, fighting in the city from house to house was a nightmare. The captain, Sergeant White and Tredanari went forward from house to house looking for an O.P. They came to a large church and decided to use it as an observation post. High up in the steeple, Sergeant White found a German B.C. scope. He set the scope up and peered through, looking for enemy movement. In the meantime, the captain, Tredanari and Pilip were on the other side of the steeple. Sergeant White stayed in the steeple to look for enemy activity.

After a few seconds, a mortar shell landed on the steeple, exploding its shrapnel. The shrapnel hit Sergeant White, as well as a few riflemen who were nearby. The captain heard the explosion and ran to see what had happened. He was in shock when he saw Sergeant White laying there half-dead. His stomach was ripped open. The captain yelled for a medic. The medic went upstairs in the steeple to give aid to Sergeant White. The medic called for more help and another medic responded to aid the riflemen.

I heard the news over my radio and told most of the cannoneers what had happened. The captain was devastated. The Krauts had an observation post in a nearby church and happened to spot our boys first and let us have it. The captain took this loss really hard. Sergeant White and Captain Smith had lived in the same town in Kentucky.

The regiment recouped and got back on the road, where we saw many of our boys lying dead in the field. We stopped and looked. We saw over 100 of our men who had given their lives for us and I recognized some of them. But there were over 500 Krauts dead there also, lying together side by side.

When the city was captured, the captain came back to the command post and had every one of us file out in a field. He gave the company the sad news about Sergeant White. At the time, the captain didn't know if he lived or died. I myself liked the sergeant. He was a soldier's soldier. I must say it again. I was fond of him. The sergeant was my hero. I missed him a lot. It has been over 50 years since his death and he is still on my mind. The memories never die. Up until this day, I still feel the pain of losing Sergeant White. When we were in Ireland, I remember some of the older dogfaces in our company would tease him and yell out, "Lauria, what color is chicken shit?" and they would yell back, "White!" It was just a joke and Sergeant would laugh it off. He was a great guy in my book.

The captain said, "I don't believe that the sergeant will make it." He then broke down and wept. After he had calmed down, he said to all of us in the Cannon Company, "If I ever see you boys take in a prisoner, I will shoot you myself." He got emotional and so did the company. I was really emotional after hearing the bad news. Chartres took a few days to capture. Our losses were heavy, and I left with some bad memories. We tried to recuperate from our losses and regain our strength. I know we were all fatigued after weeks of combat.

Life in combat was an exercise in hardship, pain and incredible stress. The enemy was not the only thing we fought against. We had the harsh conditions of rain, snow, heat and bitter cold. There were days when we had to walk through miles of mud and wade through water until our skin wrinkled like a prune. At times, our feet stayed wet inside wet combat boots the entire day before we could sit down by a fire and dry off. Most of the time our only protection against the elements was a woolen horse blanket, a woolen hat, and an army coat, which I never wore.

A sit-down meal was considered a luxury. When we could eat, we did it mostly on the move. The food came out of a can and it was cold and tasteless, and we were happy to get it. Sleep was also a luxury. We never really slept in combat. If we did, it was just a cat nap. When we were permitted to sleep, we fell asleep instantly wherever we could. Our beds were the flattest surfaces we could find. We often had to sleep out in the freezing cold rain in a foxhole. Sometimes it was so cold that the dirt was frozen solid like a rock. I slept with one blanket and coat wrapped tightly around myself with whomever else was in the foxhole to keep warm. It is very difficult to describe what combat was like to someone who has never experienced it. You never really know until you experience it yourself. We got very little sleep and even slept standing at times. We were drenched all the time from the rain and sweat. The clothes we wore stunk. They smelled of urine. Your body stinks if you don't take a bath for months. Our toilet was any place we dug a small hole with our spades. If you want to talk about being uncomfortable, this was it. Try relieving yourself with a barrage of mortar shells raining down on you and then running for cover with your pants down around your knees. This happened to me quite frequently, believe me.

There were times during the summer when I came across a small brook. In seconds, I would take my clothes off and run them through the water and put them right on again, letting the clothes dry on my body. If I had a little soap, I'd wash them a little. I did the same thing during the winter. It is a funny thing that I never did get a cold from so much hardship. With the help of God, I made it through the war.

The outfit received replacements within a few days, but not as many as we needed. Most of the time I didn't get to know the names of the new men. Many of those men would be killed or wounded, so how I do I explain the future dangers to these doughboys? Another thing that puzzled me was that many would get wounded or killed within weeks. This didn't make sense because other men would be in harm's way throughout the war and make it without a scratch. God must have played a big part in this war. I often just can't understand how faith works.

I want to explain a little about our combat shoes. We did not wear boots. It seemed to me that most of us GIs got crap to wear. The equipment was poor. The footwear in the winter really sucked. The combat shoes were not warm and water seemed to get into them when we walked in the snow or rain. They were uncomfortable. Not only did they have these canvas leggings which we had to lace almost all the way up to our knees, but they also stayed wet. The leggings were really made in the first World War—you'd think they would have come up with something better.

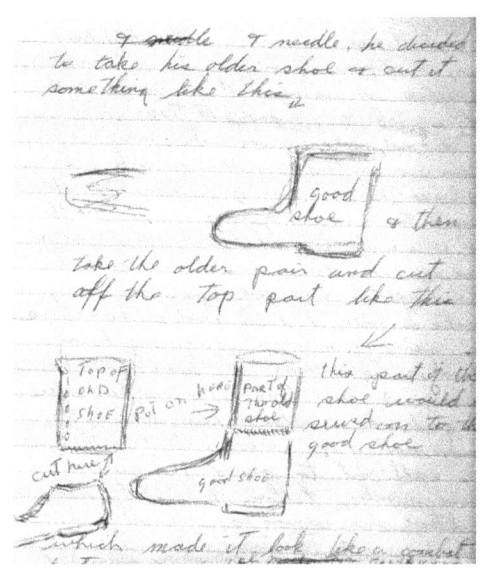

Sketch of new type of boots.

Now sometime during combat near Chartres one of my buddies, Cpl. Handerson had gotten a great idea of how to make a combat boot. Being a truck driver, he had his vehicle, where most of us carried an extra pair of shoes or O.D. He happened to have some heavy twine and a needle, so he decided to take his older shoe and cut off the top part, then sewed it onto the good shoe, which made it look like a combat boot.

I begged Cpl. Handerson to make me a pair. He had no responsibilities other than taking care of his truck, so had nothing but time on his hands. Ken Handerson took good care of his equipment and his vehicle. I gave him a lot of my cigarettes throughout the war because I didn't smoke, so one hand washed the other. I was so happy to have these new boots. I didn't have to use the leggings anymore, so they were done away with.

VI
Advancing Through France

Fontainebleau

We moved on and prepared to attack the city of Fontainebleau. I remember our cannons were on a railroad track. We began to dig in to attack the city. I was just getting ready to lay my telephone line to each gun and go forward to the new O.P. when I looked up and, lo and behold, there were five or six U.S. P-47s circling around overhead. As they passed, they tipped their wings. In the meantime, I walked alongside the railroad track and continued to lay my wire. Now the P-47s made about three or four passes over our guns. I could have hit the planes with a stone if I had tried to. They were that low, every time they made a pass.

The roar of their engines was horrifying. Our gun crews dug in while enemy machine guns fired around us. I looked up again and here came another P-47 straight at us. It was flying parallel to and straight over the railroad tracks, heading straight at me. This time the plane was about two hundred feet up. All of a sudden, the plane dropped his egg. I first thought it was his reserve gas tank. It came straight down at me.

As I looked up, I realized that this was not a gas tank but a 250-pound bomb. I saw it coming almost right at me. I dropped my D.R.-8 reel and made a desperate leap off the embankment. The train tracks were about ten feet high. As I went down, the bomb exploded on the opposite side of the railroad tracks. The concussion was so strong that as I jumped down it slammed me against a tree. My legs landed wide open and I hurt my groin. I lay there for some time and could not move. Everyone else had run for cover.

I was in so much pain I wanted to cry. I heard some of our officers yelling, "Put out the panels! Put out the panels!" The panels were made of shiny material and were supposed to be on all vehicles when we advanced. It was hard for American fighter planes to know who was on

Advancing through France.

the ground, the enemy or American doughboys. Thank God the bomb landed away from the six guns. That would have been a terrible accident. The whole Cannon Company would have been wiped out. After we put the panels on our vehicles, the P-47 airmen recognized us and came around again, tipped their wings, and left the area.

Seine River

I was still in pain for a good hour. Someone else took my place and went forward with my wire. After a few hours, I was sent up to the O.P. and resumed my job. We were in Montereau by the Seine River and prepared to make a crossing. When my outfit tried to attempt to cross the Seine River at Montereau, Captain Smith, Tredanari and I hid behind some tires stacked four to five feet high. We were about 40 feet up on high ground. Some of our riflemen were on a raft which was constructed by our engineers. There was a three-quarter-ton truck on the raft with about a dozen riflemen on the near side of the river. The river must have

72 Running Wire at the Front Lines

An American P-47 dropped his egg on our position as I was laying the telephone line to the guns. The concussion was so strong that as I jumped down it slammed me against a tree. My legs landed wide open and I hurt my groin. I lay there for some time and could not move. Everyone else had run for cover.

been three or four hundred feet wide. The enemy shot at the riflemen on the raft while we were on the high ground behind the stack of tires, returning fire.

We hit the far side of the bank with our shells to keep the Krauts from getting accurate shots at our men on the assault boats. They tried to get a vehicle across on a raft. Engineers got ready to build an assault bridge as soon as the riflemen got a foothold on the far side of the river. In the meantime, our O.P. crew sent more fire missions to the guns. We had to keep a constant barrage of shells exploding to keep the enemy in their foxholes or whatever they used for protection. This went on for a good five hours. The doughboys made progress crossing over to the far side of the river. The enemy continued to hit us with artillery and machine-gun fire. I can still see myself behind a pile of tires about four feet high.

The captain told me to take a break so I left my phone with someone in our crew and walked over to a barn that was about ten feet away. I walked in and to my surprise there were ten or more doughboys sitting on the ground or on boxes and drinking hard cider. There must have been about ten big barrels. They were big enough to walk in and could hold about five hundred gallons of cider.

I saw someone I knew from back home. He was my dentist and a major. He recognized me and asked me how I was doing. He told me to sit down and have some canteen cups of cider, which I did. I wasn't much of a drinker, but I sat and talked to him for a short time and started to get a little dizzy from the cider. I told the major that I had to get back outside to my post. He offered to make me a new partial plate for my teeth after the war was over. I thanked him and went back to my post outside the barn.

Little did I know that within a week, I would find him dead with his head cut off from artillery shrapnel. This was the third doughboy I knew that was decapitated by shrapnel. About a week after we had the break in the barn, our unit came to a halt on the outskirts of a French town. There was a large field and we were shocked to see so many dead men in it. Most of our boys jumped off the truck to investigate what happened. Some of the boys and I walked through rows of dead soldiers. Around 50 or so belonged to our unit and over 250 were Germans. I saw one soldier whose head was decapitated and nowhere to be found. I looked on the shoulder pad and it was the major who I had a drink of cider with just a week before. I was in shock. I was positive it was the major. He was a large fellow and easy to recognize. I fell to my knees and made the sign of the cross, got up, and walked back to the wire truck with tears running down my face.

A week earlier I left the men drinking in the barn and returned to the O.P., where we continued to shell the far side of the Seine River and made progress. Our doughboys were getting a foothold on the far side as long as we kept pounding the enemy with mortar and artillery shells, giving our engineers time to build an assault bridge. More and more doughboys got across and the bridge was almost complete. The captain told us to get ready to cross over to the other side. We crossed the river and got into the city of Fontainebleau.

As usual there was street and mouse-hole fighting. We went from house to house and rooted out the Germans. This went on for a few days. Then they fled to a new position. At the beginning of the break, they tried to form a new front line. They were so disorganized.

Thank God the Germans left their position to form a new line of defense. While occupying the town, we waited for orders to see where our unit was supposed to go next. In the town, French people celebrated and gave us wine and cider. One lady gave me a bunch of flowers that she picked from the fields. When I looked inside the bouquet, I saw a four-leaf clover. I had never seen one before in my life. I picked it out and put it in my wallet. I kept it for over 25 years, until it disintegrated.

The French people in the town danced and kissed the soldiers. After an hour or so, I heard some screaming. FFI Frenchmen were dragging some women into the town square. The women tried to resist the Frenchmen and kept screaming. People in front of me were laughing and clapping. I didn't know what was going on. I looked above the men's heads and saw the Frenchmen cutting the women's hair off down to the scalp.

The women yelled and screamed in protest. The Frenchmen cut the women's hair because the women had collaborated with the Germans. They did this to eight or nine women to shame them. This was how they were punished.

Reims

Somewhere in France, when the outfit traveled through the country, I remember fighting at a town called Reims. It was a well-known city. On our way there we went through the outskirts of Paris. I can still picture in my mind the church called Notre Dame, in Paris. We could see it from about ten miles out. Most of the land in that area was flat and good tank country. That was the closest we came to Paris.

Our regiment battalions captured a lot of Germans in that area. My captain told us to line them up in a straight line, shoulder to shoulder, so he could interrogate them. He walked up and down the line and asked each prisoner what his nationality was. All of the prisoners had German uniforms on, but the captain got an unbelievable variety of answers. Some of them claimed they were French, Polish, or Russian. I stood nearby and heard all of their answers and said to a nearby doughboy, "Who in hell are we fighting?"

The captain then asked a prisoner if he was German; he denied it and said he was Polish. The captain asked him again, but he denied his nationality once more. This went on a couple more times. The prisoner next to him kept insisting that the prisoner under interrogation was a

German. Finally, the frustrated prisoner went behind the prisoner under question and kicked him in the ass three times, repeating, "Deutsch, Deutsch, Deutsch!" After that, the prisoner admitted that he was in fact a German and that there were a few more in the line. The captain ordered the GIs to take the prisoners for more interrogation. Most of the GIs laughed about it.

I spoke up and wanted to know what was going on. Were we fighting everyone else but the Germans? I suppose the Krauts were putting all the captured men from other countries in uniform and forcing them to fight. This was against the Geneva Convention. When the Germans took over a country, they got all the able-bodied men and put them in uniform or the work force. Most of the time, when we captured the enemy and prisoners came out of hiding, the men yelled, "don't shoot" and "nix Deutsch," pleading with us not to shoot them because they were Polish, French or Russian. After a while, the number of people of other nationalities that we encountered when we caught "German" prisoners died down.

After Verdun

My company took up a position in a wide-open field which looked like the hedgerow of Normandy. The area was right outside of Verdun, a French town we had just captured. Up the road, something strange was going on. I looked through a gap in the hedgerow that was wide enough for a tank or a truck to go through and saw about six Frenchmen who were part of the FFI. These men kept arguing and pointing down the road. I got curious and went to see what all the commotion was all about.

I looked in the direction they pointed and lo and behold I saw German movement. The Germans were about two hundred yards down the road, setting up a 40-millimeter gun and mortars, waiting for an American tank to come down the road. The Frenchmen wanted to capture or kill them. I yelled at the Frenchmen to get off the road and conceal themselves, but they didn't pay any attention to me.

In the meantime, I ran back to our Cannon Company and called for the executive officer and told him what I had seen. The officer ordered one cannon to move up and get a direct hit on their position. For some reason, we did not have any riflemen in our company then and the cannons were our front line. The cannoneers were pulling the cannon to the road. They still had about 50 feet to pull the gun.

I looked and saw the Frenchmen still arguing out there; the next thing I heard was a huge explosion. The Germans had spotted the French FFI and fired the 40-millimeter and mortars on them. All six of them died instantly. After that, whenever I saw FFI Frenchmen I went the other way. The French seemed to fight with their hearts rather than their heads. Don't get me wrong, though, during combat in France, the French gave us a welcome greeting whenever we liberated their towns. They were also helpful about giving us information. They gave us flowers and wine or cider. When we took a small city or town, the French were really helpful about telling us where the Germans were. They were our eyes and ears at times.

VII

Dornot

In early September, the 11th Infantry Regiment came to a halt near the French town of Dornot, which was just before the city of Metz. Our unit had just finished clearing out a few towns and moved ahead. Fighting was dreadful but the unit forged on, forcing the enemy from one town to the next. As we approached Metz, there was lots of talk about how fortified that city was. It was September 8, 1944. Our regiment, plus all of its units, got into position. Our company guns were in a large field facing east which must have been a farm because it looked to me like it had just been plowed under for next year's cultivating. On the north side of the field was a wooded area running over six or seven hundred yards.

When we got to this position, I started to run a telephone line to the six guns and then over to the command post. There was always an executive officer in the C.P. It could be an open field or a house. This time we used a house that was built over a bunker from World War I. The farmhouse was the same length as the bunker. It was really deep. French farmers always seemed to build their houses in a square, boxed in with other buildings and a barn. It was a way they would be protected from outsiders. The bunker was used as a cellar for wine and storage but I didn't go down there. I ran the wire to the guns and back to the house, and went up into the attic because Steve told me to meet him up there after he had laid the wire from the bunker to the attic.

When I was out by the guns setting up the telephone, there were four or five Sherman tanks running back and forth along the wooded area, shooting their machine guns into the woods to flush out some Krauts. The Sherman tanks made a huge racket. I knew for sure the Germans would send mortar or artillery fire down on our necks real soon.

I gave this some thought but started back to the house. Most of the gun crew dug in, making things a little safer. I went back into the house and went up to see Steve, who was on his Radio 300. Communication

HOUSE BUILT OVER A WOULD WAR ONE BUNKER FIFTY OR SIXTY FEET long

Our location on September 8, 1944, in a house built over a World War I bunker that was over 50 feet long. Our radio antenna is poking out of the tile in the roof. James Stephens was with the second cannon when he was hit with German shells and killed. He was just 19 when he died. He grew up in Wisner, Louisiana. In 2001, I shared details with his younger sister and niece about what he experienced in combat and how he died.

coming from the observation post was poor and Steve said, "Lauria get your carbine and poke one of the roof tiles out and we'll put the radio antenna up into the hole in the roof for better reception." I did and the reception was much better.

As we talked, a few shells came near our position. There was a small window upstairs and we looked out to see what was going on. Some of the shells exploded but were not too close to the guns yet. I was worried because the attic had no protection. It was just one open room with no walls. It was flimsy and it only took me a second to poke out the roof tile. Steve told Ted at the O.P. that we were getting shelled over by the guns. Ted told Steve that they didn't see any German guns firing at us and the conversation ended.

About five minutes went by, and lo and behold a barrage of more than 20 rounds came in on us. They exploded within ten yards of each other. The noise overhead made it seem like they would come directly into the attic. We had a good idea how close to us shells would explode after two months of combat. I didn't like what was going on; taking chances up in this attic was suicide. I told Steve, "Let's move downstairs somewhere." He said that wouldn't work because the radio reception was too poor. So we stayed there.

A few more minutes went by. I could still hear the tanks running around but couldn't see them because they were on the north side of the barn, near the guns. The first barrage of shells landed on the northeast side of the house and we could see the shells land and explode out of a small window. After that, the Germans adjusted their guns and were now on target. The only good thing about the barn was that it had good cement walls; otherwise the inside was an empty shell with no upper protection. The stairs were all open. If a shell went into that roof, it would explode on the ground floor and kill about 20 of our riflemen who had taken cover inside the barn.

Another barrage of shells came in around the courtyard, which had an open stable for the livestock. I told Steve we should get out of there and once again he told me no. We both waited for a few minutes and Steve called O.P. There was no sign of the enemy. Once again another barrage came in on us and the shells landed and exploded in the courtyard. The whole area seemed like it was blowing apart. I dropped my telephone and ran down the stairs to the ground floor. Steve yelled for me to stay but I didn't.

I waited and made an attempt to go back upstairs. I got halfway up the stairs and yet another barrage came in. I headed down to the bottom of the stairs and tried to make it outside the barn and around the house to the bunker. It was too dangerous to attempt so I had to wait. In the meantime, all the dogfaces were sweating out this artillery barrage. There was a large front door on the barn and next to the door was a large window about 8 feet high and 12 feet long with small panes of glass about six by eight inches.

Sergeant Turner just happened to be taking cover inside the barn. He saw me pacing back and forth, wanting to dash out the door. He called me to come over by him and asked me what I was doing. I explained that I had left my post upstairs and he told me to stay put. He knew I was panicking. He heard another barrage of shells coming in and he grabbed me with two arms and held me down on the ground. This barrage was worse than the last one. I wanted to dash down to the bunker to safety but the sergeant told me to stay put. While we held our ground, over 60 shells landed in this small area.

On the last barrage, Steve ran down the stairs. There were 20 GIs huddled up there on the ground floor. There was a lull in the air. I didn't listen to my sergeant and got ready to run out into the bunker. I started for the door but Sergeant Turner saw me and grabbed me just as another barrage of ten or more shells came in. Some landed in the courtyard. He pinned me to the ground. As I fell on my face in the direction of the large window, the barrage hit the front door and all of the window glass blew out and hit me in the face. Sergeant Turner held me down again. In this war, he was my angel of mercy.

I waited for the right moment to go down into the bunker. The problem was that I had to go out the door and go around the corner of the barn and run another 15 feet to the bunker steps. It seemed like a long way to go. If the Germans decided to hit us again while I ran, I would meet my maker. Sergeant Turner told me to stay put. Steve wanted me to go upstairs again but Turner said, "No, stay with me."

Lots of shells landed inside the courtyard, which wasn't that large. It was about 50 by 50 feet and the cobblestones made it even more dangerous to run across. Steve came down the stairway. I couldn't take much more of this. I started for the doorway which had been blasted off and ran for my life. The sergeant told me not to go but I didn't listen. I made my move and went. I made it to the bunker steps and to safety.

To my surprise there were over 50 women and children down in the bunker crying and praying. Some must have crapped in their underwear. I could smell it. The sound of praying made it seem like I was in church. Others used rosary beads to pray. The shelling started up again. Some medics brought in the wounded. Two of them were my buddies Stephens and Jambois. Stephens was hit and had two medics dragging him in. Jambois was walked in with a concussion. Some of the boys sat him down.

Stephens had been with the cannons when he was hit by the shells. The medics rolled Stephens over on his stomach. He was moving all around and the medics tried to keep him still. He kept moving his arms. They lifted his shirt and gave him a few shots of morphine. I looked down at him and saw a cut over eight inches long by his kidney. The blood was coming out like a garden hose. I couldn't endure looking. Our executive officer yelled out for a jeep driver to take Stephens to the field hospital.

In the meantime, more shells came in while I was in the bunker. Lt.

Schaech, my officer, yelled out, "Who is upstairs on the phone?" Lt. Schaech got on the phone and tried to reach me and there was no answer. He yelled out again, "Who is on the phone up in the attic?" I said, "No one, sir." He then yelled at me, "God damn it, Lauria! Get your ass back up in that attic before I have you court-martialed!" I said, "Yes, sir!" and ran as fast as I could. Steve was down on the first floor. I said to Steve, "Let's get upstairs before Lt. Schaech has us court-martialed."

When we got back up to the radio and phone, I called the bunker to let the lieutenant know we were at our post. I told Steve how nice and safe it was in the bunker, but for us it wasn't going well. Steve and I were upstairs like sitting ducks. The shelling had intensified. A few minutes went by and Steve got a call from Tredanari at the O.P. calling for a fire mission. They spotted the Krauts and guns that had caused so much damage to our company. I called down to Lt. Schaech. Steve talked loudly so I could repeat the information to the lieutenant to get it to him faster.

Within a few minutes, I got the reading from Lt. Schaech and repeated it to the guns. The number one gun got the job of shooting out the first round. The call came back from Tredanari that the shell was about a hundred yards short and about a hundred yards wide. I got on the phone and told the lieutenant. He gave me a new reading. I repeated the new reading to the guns. This time the lieutenant told me to have all guns adjust to the number one gun as soon as they were ready. I called and all guns were ready. I repeated it to the lieutenant and he told me to have the guns fire. I executed the order to the guns. All rounds went off. We waited a few seconds and the word came back from O.P. that they were on target and to fire at will. I told Lt. Schaech that the captain wanted all guns to fire at will, and so he had me repeat the order to the guns.

After a minute, Tredanari told us to cease fire. About 60 or more rounds had been fired. Steve asked Tredanari how things looked. He told Steve the enemy had either been killed or dispersed and scattered, leaving their weapons behind. The enemy was no longer in the area. The captain got on the radio and told us job well done, the enemy guns were knocked out of action. Then Steve told Tredanari to tell the captain what had happened to the company. Steve and I didn't know the full story yet. After about an hour, I was told to take a break with Steve.

We went down to the courtyard and saw all the damage that had been inflicted on us. Some of the new men were dead in the open stable. I had only known them for a few days. I saw all the shell holes made by the enemy guns. I walked over by the guns and asked who was hit. The cannoneers told me some 20 or so boys were hit and rushed to the aid station. The gun crew licked their wounds. I guessed we would get some new replacements in a day or two.

I talked with the cannoneers, got the names who was hit and stayed for a short time. They told me who was hurt more than the others. It was sad. After a few days, most of them were back in the company. While I walked away from the gun crew, I had tears in my eyes. I just couldn't get over this nightmare. Most of the men spent over a year with me in Ireland. I cried with anger while walking away, then went back to the farmhouse with red eyes. The only good thing about this was that we killed most of the Krauts that had done this to us.

I am sure that one of the Germans had a spotter or O.P. that was directing this fire mission. That fire mission was well done. Their concentration was on target, not firing at random. It was well planned. I went back inside the barn and asked about Pfc. Stephens when our jeep driver came back to the company. He said it didn't look good for Stephens. I felt awful seeing him bleed so much, with that eight-inch shrapnel inside his kidney. I am sure he died on the way to the aid station. I never heard from him again.

Some days I couldn't cope with all that was going on. Day after day someone got wounded or was close to death. This war was getting to me. There were times when I stuck my head out and did brave things and some times when I couldn't cope with it. I guess there are limitations to how much a man can take. This happened on September 8, 1944, at about 1000 hours.

During the attack, a shell had landed a foot or so away from one of my buddies named Jambois, a real nice guy, maybe in his late twenties, who was in his foxhole. He was walked into the bunker and was taken to the field hospital. In a few days, he was sent back to the company. Then he was sent back again to the hospital. This went on for a week or two. Jambois was in shock. If anyone talked to him, he gave no answer. We kept him with us in the command post.

The captain asked one of his officers why was he still there and not being taken care of in a hospital. They explained that the hospital sent him back to the company. Captain Smith got on the phone to the regiment and told whoever was in charge to have Jambois looked after. If not, he would go to the field hospital personally and put whoever it was that sent him back up in the front lines with the rest of us. The captain had lots of pull so he got some results after that phone call. After that incident, Jambois was taken care of. We never did hear from him again, but I am sure he was sent home safe.

For the rest of the day, I repaired some wire that went out to the guns that was damaged during the German shelling. I was upset most of the day with our executive officer, Lt. Schaech. He was always arrogant but later on in the war he lost his privileges of being our executive officer because he was even arrogant with Captain Smith.

After that bad day at the barn and the intensified shelling, we tried to lick our wounds. We stayed sad the whole day. After repairing my wire to the guns, I went indoors. The command post in the bunker called for Sergeant Turner to come down. Captain Smith arrived back from the O.P. with Tredanari and Pilip and met with Sergeant Turner in the bunker. Sergeant Turner came back to our room and told us that the captain wanted him to get a few radio wire men to go forward that night and set up a new observation post for the next day's attack. Turner picked me and Godzisz, Campbell, Pilip and our wire-truck driver, Pfc. Turner. Steve remained back at the command post.

We put most of our equipment on the wire truck and set out for the new O.P. The truck was on the south side of the house. There was a narrow dirt road running east directly to the front lines. Captain Smith gave Sergeant Turner all the details on what we should encounter, so we brought enough wire, equipment, guns and rifles. We started out. Our truck driver had to be careful because the road was so narrow. I laid the wire alongside the road and made sure the line went by the edge of the hedgerow. I made this a must because if a tank went down this road it would destroy our phone lines.

We were about two hundred yards down the road when a doughboy halted us, so we gave him our password. He replied with his password and then asked Sergeant Turner where we were going. Sergeant Turner told the guard we were going up front to set up an observation post for the morning attack. The guard said, "You guys are crazy, this is the end of the line. Once you go past me, you're on your own." This was the last outpost. He said, "Good luck boys." and we all said "Thanks." So Sergeant Turner told Pfc. Turner to drive his truck about a hundred yards to our left. We took our equipment off the truck. Pfc. Turner was to stay with his wire-truck. This was a good move because the truck driver had some protection from the guard. There was a dense wooded area near where the truck was parked on the left side of the road.

When we started out, the sky was cloudy and really dark. In front of us was open, hilly land. As we got up to the first hill, another hill began. We continued going up and our old reel emptied so we spliced a new reel to the old line. We had a reel called a D.R. 4 which holds almost a mile of heavy wire on it. Sarge told us all to work quietly and stay close. As we started out, the reel made a squealing noise. I said to Godzisz, who carried the 300 radio on his back, that I wished I had some oil to put on the shaft.

We were all nervous but walked up hill at a slow pace. As we kept going up, I noticed the wire on the reel was half finished. I told Godzisz to tell the sergeant we should stop because the captain said the outpost wasn't too far from the last outpost. The captain told sergeant that he would know where the spot was. I told Turner that we had used a half mile of wire since we left the guard at the outpost. Sergeant asked Pilip, "What shall we do?" He talked it over with Sarge and decided to go another hundred yards and stop.

The boys and I were very nervous. The reel was making a racket. Campbell and I put down the reel. It was almost empty. Turner talked it over with Pilip and they decided to leave the radio and wire crew in place while they went forward a few hundred yards. Godzisz, Campbell and I stayed put.

When we started around midnight, the weather was cool and there was no moon. As we waited for the sergeant to return, the clouds started to break up and the moon made it too bright. I said to Godzisz, "I don't like this. It is getting too light." We were about ten feet from the top of a high hill and could easily be spotted, so the three of us stayed hugged to the ground. Then Godzisz said, "Lauria, we are going to call Steve back at command post with the 300 radio." He tried a few times but didn't get a response. Godzisz tried again and still got no reply. He tried to get him about ten times. Godzisz started to call and speak in Polish to Steve—they were both Polish Americans—but there was still no answer.

Now Godzisz started to swear in Polish. I told Godzisz to speak low because the Krauts may hear us. We agreed that Steve must have fallen asleep at his post. The three of us were getting worried because sergeant and Pilip were not back yet. It had been over a half hour. Twenty more minutes went by. The wind blew harder and since we were so high in the hills it made a lot of noise, especially with a helmet on. Godzisz called back to the command post again and still had no luck. We were getting nervous waiting up on the hill. We stayed as quiet as possible.

I heard voices, but shook it off and told myself it must be the wind. Again, I heard voices. I whispered to Cpl. Godzisz and Campbell to listen because to me it sounded like Germans talking and digging. They told me it was all in my mind and it was the wind. I tried to convince them again and so they listened. This time they told me, "You are right, Lauria. There are Krauts and about 20 feet away." They must have just gotten there and started to dig for the morning attack. The best thing for us to do was to move away from the location and look for Sergeant Turner and Pilip

We started to move to the right on our hands and knees to put some distance between us and the Germans, leaving our equipment behind. If we saw Sergeant Turner and Pilip, we decided we would stand up and yell for them to run and move to their left away from the Germans. We lay still for about five minutes. This seemed like a lifetime. I prayed for them to show up. I had exceptionally good eyesight at that age. I looked in the direction that the men had started in. I kept looking and so did the other boys. I looked and looked and didn't want to give up hope. The strong wind made my eyes tear. I would dry them and then look again and pray for them to return. A good half hour went by when, thank God! I saw two objects in the distance. I told Godzisz that I saw sergeant and Pilip at a distance but wasn't completely sure. The objects were too small to make out. Godzisz told me I was seeing things. I waited a few seconds and the objects became larger. Godzisz and Campbell said, "You are right, Lauria. The two dots are getting bigger."

I told Godzisz and Campbell, "If it is them, we should wait until they are about 50 yards away to jump up and yell for them to run to their left away from the Krauts." I whispered to Godzisz and Campbell when it was time to count, "One, two, three, all together as loud as we can." The wind was in our favor. We jumped up and started to yell. Both of the guys understood what was going on. The sergeant and Pilip yelled for us to drop our equipment and run for our lives.

Little did he know, we had already left everything by the Germans about 75 yards to our left. The only thing we kept with us was the radio strapped to Godzisz's back, our carbine and the DR-4 shaft. As we ran down the steep hill, which was about a thousand yards long, I was in the lead. My leaps were at least ten feet long going down. I bounded like a deer. I heard the Krauts yelling in German behind us "Achtung! Achtung!"

Since I was in the lead, I called to Pfc. Turner to turn the wire truck

I told Cpl. Godzisz and Campbell, "If it is them, we should wait until they are about 50 yards away to jump up and yell for them to run to their left away from the Krauts." I whispered to Godzisz and Campbell when it was time to count, "One, two, three, all together as loud as we can." The wind was in our favor. We jumped up and started to yell. Both of guys understood what was going on. The sergeant and Pilip yelled for us to drop our equipment and run for our lives.

around. I guessed he had some idea that something had gone wrong. I had about 70 yards to go when Turner started up the truck and tried to turn it around. The damn truck's breaks squealed like crazy. They made a hell of a sound. We hopped in the truck and Sergeant Turner told our driver to make it snappy. The brakes were noisy but Pfc. Turner got the truck straight on the road and started to speed away.

Mortar shells rained in on us. Thank God they all missed. I was pissing my pants. The doughboy at the outpost said, "What in hell did you guys do up there?" The sergeant told him what had happened. The guard told the sergeant that he was almost hit with a shell. He told us to go back to our command post and stop starting a war. We got back to the C.P. about three in the morning. The sergeant told us to get some sleep. In the meantime, Godzisz went looking for Steve, who was asleep.

When he found him, he gave him a hard kick in the ass for falling asleep on his post. Godzisz called him a few choice words in Polish and then bunked down to get some sleep. Morning was just few hours away. We would get up soon.

In the meantime, the sergeant went next door and let the officer in charge know what went wrong at the new observation post. In the morning, we would attack. If we succeeded, we would close in on the city of Metz. This city had about 26 forts surrounding it. Little did we know, it was going to be our bloodiest battle in this campaign.

We were all up at 0500. Those of us who were up most of the night were up again with just two hours of sleep. I was awakened and told to get ready to move out for C.S.M.O., or close station march order. The captain got the news about what had happened with the wire crew during the night. He called for Sergeant Turner and wanted an explanation. Sergeant Turner told him the truth about what had happened at the new O.P. but the captain called him a liar and insinuated we didn't go where we were told. Sergeant Turner told the captain that when we started the attack he would take the wire in his hand and show him the equipment that we left exactly where we were instructed to go.

Sergeant came back to the radio and wire men. We were waiting by the side of the barn for him. He was pissed. He told the crew what the captain had said, insulting the men who were on the assignment during the night. It was still a little dark, but light enough to see that everyone was ready. The attack kicked off. The men who were up that night stayed with the sergeant and Pilip. Most of the time, Pilip stayed with the captain and radio operator Tredanari.

The morning attack began. As we went up the hill, the Germans spotted the riflemen and started to fire on them. Mortar shells came in on us. It was not a piece of cake to take high ground. Our cannoneers were told to fire at will until we got to the high part of the hill. Then we would have the advantage. When we got to the high ground, Sergeant Turner still had the wire in his hand that we had laid the night before. I couldn't believe we had gone that far yesterday.

Sergeant called Captain Smith and showed him the equipment Campbell and I left behind. We were the ones that were responsible. Captain apologized to us and said, "My God! You boys went deep into enemy lines." He was amazed that we had gotten out of there. He said, "You boys are lucky to be here. You could have been captured or killed." He apologized again.

Captain Smith was a good officer. I spent time with him at the O.P. alongside Tredanari and Pilip and when we passed each other in a field or spent some time at our command post he would always say, "How are you doing, Lauria?" I would say, "Fine, sir" and always have a smile for him. He wasn't a tall man, only about five eight or nine. I spent many hours with him inside the command post when he was there or up front at the O.P.

There was one thing he always got on my ass for. I had a habit of letting my helmet strap hang down. He would yell, "Lauria put that damn strap back around your helmet and don't let me see it hanging down again." The reason I kept making it hang was that I was superstitious. When we left the ship and landed in Normandy, all of the GIs had orders to put their helmet straps up around their helmet. That was a strict order. At first, we could let our strap hang or put it under our chin. So, when we landed in Normandy, my strap was hanging down, and after that I never wanted to put it back up where it belonged. I was afraid bad luck would come to me.

We had orders to wrap the strap around our helmet because with the strap under your chin the Krauts could sneak up behind you, pull back on your head and choke you with it. When the captain saw me with the strap down, he would give me hell, but the minute he was out of sight I would put the strap down again. This went on all throughout the war. Dornot started on September 8, 1944. It lasted until September 10, 1944. We had another eight months to go.

VIII
Metz

Metz was a large city and had over 25 forts. It was considered one of the most fortified cities in all of Europe. As we approached Metz, we had to cross the Moselle River. The Fifth didn't wait for orders to cross. We believed that when the opportunity was there we should take advantage of it. We crossed the Moselle River at Dornot. It was raining and the weather was miserable.

The infantry combat team crawled slowly down the road. The Cannon Company took up position on a hill named 334. The hill was high enough to see for miles around. In the meantime, the Second Battalion, 11th Infantry, and some of K Company were fighting hard. Some of the doughboys carried an assault boat down to the Moselle River.

The Cannon Company was prepared to hit the enemy as soon as the O.P. located any enemy movement. I ran my line down the hill, which was full of old shells, left there since the First World War. The hill was part of the Maginot line and old trenches remained for miles and miles. They were rebuilt and made stronger.

Every hundred yards or so there was a concrete bunker. I slept in one for over a month. It was cold and damp. When it rained, water dripped into it from the top because the roof was flat and the cement porous. The bunker was about 10 feet wide and 20 feet long. At times, it rained for days and there would be two or three inches of water inside. The bunker was our command post, with the cannons about six hundred yards behind us. I'm not complaining about the living conditions. To me, it was a safe place to be since the bunker was made of reinforced concrete and steel. With all the mortar and artillery firing at us, it was worth being in the bunker.

The Second Battalion command post was located in the town of Dornot and had been under fire continuously. The command post had been hit with a barrage of shells that knocked out the battalion radio and

Walking in the World War I trenches outside Metz.

killed three men and wounded several others, including the executive officer and G-2 captain. The next day the enemy hit the command post again and it was forced to move to a new position.

Our command post was moved to the west side of hill 334. We were constantly exposed to artillery and mortar fire. We kept the cannons so far back so that it was hard for the Krauts to locate them. Down below us was the town of Dornot. There was a lot of activity going on. One good part about the bunker was the officer was able to use a fold-away table to put his map on to calculate the fire mission. Using the table was much better than having to use the wet ground.

Steve and I were in the bunker with our executive officer Schaech. Steve received fire missions from the O.P., and Lt. Schaech kept busy with map readings and his slide rulers to calculate readings on the map. The moment he gave me information, I repeated it to our gun crew. Once they told me that the guns were ready, I told the executive officer and he gave me the order to fire.

Well anyway, the day was a really busy one. We knocked out a few tanks and also broke up some troop movement. Whenever the guns hit

This is somewhere outside of Metz after we found some eggs in a barn. We made eggs, corn beef hash and some coffee.

a good target, the captain told us or the lieutenant the good news. After a while, the fire missions stopped. The gun crews had worked their butts off almost the full day. Everyone took a break.

After about a half hour, the O.P. requested a new fire mission that only involved the number one gun. Now for an hour or more the mission didn't sound right. The reading seemed to be going from one end of the field to the other. Lt. Schaech asked Steve and me, "What the hell is going on at the O.P.?" Lt. Schaech was getting really angry but still did his job.

I gave the number one gun a new fire mission. They stood by for the command to fire the gun. Lt. Schaech hadn't given me the orders yet to have the cannon fire, and by mistake, I executed the command to fire the number one gun. Lt. Schaech jumped up with both hands in the air and yelled out to me, "God damn it, Lauria! Who gave you orders to execute the command?" I said I was sorry for what I did. It was a mistake. He kept going on and on for a good five minutes. I felt awful about it. Steve put his head down, trying not to hear how bad the lieutenant was carrying

I got to see General Patton when he visited our division sometime during the battle of Metz.

on. Schaech got on the radio and asked the captain what the hell he was doing up there with the calculated fire missions. He was really nasty with the captain. The captain himself told him to get his ass out of the command post and he was relieved from his duty. The lieutenant was sent back to the guns as a platoon officer.

The captain told me to get Lt. Roggenstein to come up to the command post and take over the executive officer's duty. I went back to the bring Lt. Rocky with me because he didn't know where we were located. We walked through the old World War I trenches. When we arrived, he went to speak to the captain. The captain told him he was the new executive officer and made him a first lieutenant. That was all he was told.

The next day, Lt. Rocky asked Steve and me what had happened. So Steve told the lieutenant that during the fire mission the day before Lt. Schaech had had an argument with the captain. Lt. Schaech told the captain that he didn't know what he was doing, giving orders for the number one gun to shoot from one end of the area to the extreme other end, which is called traversing. That's when the Captain had him dismissed. Steve also told Lt. Rocky that before he lost command as executive officer, Lt. Schaech gave Lauria a really hard time, yelling at him and calling him stupid and everything under the sun. Lt. Rocky shook his head and said he was an ass. Lt. Rocky had little love for him anyway, like most of our crew. The only one that suffered with him was Steve. Steve was his radio man and whenever the Second Battalion went into an attack, Lt. Schaech and Steve would go forward in support of that battalion. So now, Lt. Schaech was second lieutenant and in charge of Second Platoon in the Cannon Company.

The fighting went on day in and day out. It rained constantly and everyone was wet from top to bottom. The 10th and 11th Infantries prepared to make the river crossing of the Moselle and the Second was in reserve. With all the rain we had gotten, the river was swift. The current from so much rain wasn't going to help the GIs survive this crossing. The doughboys carried their assault boats down to the riverbanks. It was about six bells, not quite light, but safe since the enemy couldn't see them. This was our element of surprise. The Cannon Company and the rest of the Fifth Division artillery were to hold off hitting the far side of the river so our doughboys could get across without being detected.

Things started off well but didn't last. The doughboys got a foothold on the far side. For some reason, the Krauts waited for them to get in deep and let many of our troops cross over before they opened up. Now all the German forts opened up on our rifle company. Krauts had surrounding forts all around our riflemen. They fired on our troops and many of the men got hit. Cannon Company got a fire mission. The O.P. had spotted enemy guns. I got the call and with no time to waste, I called the guns for a fire mission. Things were going to get better.

As far as the executive officer goes, Lt. Rocky was a wiz when it came to calculating numbers. He was a graduate of Notre Dame and didn't need a slide ruler. He figured out everything in his head. Rounds went out so fast and the shells hit just where they were called for. This put the Krauts back in their holes and gave our riflemen a break to recover as the artillery and mortar kept pulverizing the Germans' positions. This went on for a good two hours.

Our doughboys had to pull back a few hundred yards. The problem was that the enemy had all the forts on the high ground and could see almost all our movement. The forts kept our riflemen in disarray. The battalion had orders to pull back and let all artillery and mortar fire into enemy lines. Cannon Company did likewise.

Enemy aircraft were now in the air and concentrated on bombing our assault bridge. They were successful and blew it up. Most of the two battalions were dug in on the far side of the river and fighting for their lives. The artillery and mortar fire were no match for the heavily defended German forts, with their heavy artillery weapons and reinforced concrete bunkers. The forts had tunnels going between them. Some even went into the city of Metz, where they got their supplies.

The regiment called army air force for air support. Within an hour, we had help. P-47s came diving down on the forts, dropping their eggs on the turret guns. I remember this became routine. The P-47s would show up every morning, noon and night, dropping bombs on the forts. This kept the Germans from doing damage to the riflemen. If it had been peacetime, I would have wanted to see a show like this. You could not pay for it. There were around 36 planes that flew in and did their job, dropping 250-pound bombs. That made it around 216 bombs a day. The funny thing was that the planes didn't make a dent on the forts. This was a tough situation.

We did have other bridges in our sector but they had been hit with heavy artillery and mortar, taking them out of action. In the meantime, the word was sent out that we were out of gas. There was only about two thousand gallons of fuel within the entire Fifth Infantry Division. The GIs were discouraged by the slow progress of the advance and from being hit from all sides. The doughboys were cold, drenched with rain and living in bad conditions.

After a week of this warfare, orders were sent to all company commanders to try to retreat across the riverbank to the west side. During the night, GIs had to make it to the river bank and try with whatever was available to get across and retreat. Some couldn't swim. Some tried to get across with ropes that were tied from the west bank to the east bank. Some of the wounded were put in assault boats and made it back to the west bank. Others, who tried to swim, drowned halfway across from exhaustion. Those who fell asleep in their foxholes didn't know what was going on and were left behind.

I remember during the retreat across the Moselle River, one lieutenant swam across the river over ten times. Each time he came back with one of his men. He went out one too many times and he was never found again. I could go on and on about other GIs who put their lives on the line for their buddies. I saw one rifleman with a machine gun holding off a platoon of enemy soldiers. Most them lay dead just a few feet from his gun before he himself was killed. Sometimes in an assault crossing when a doughboy was drowning in the water another would save him under heavy mortar and machine-gun fire. Other times foot soldiers would carry the wounded out of harm's way. I witnessed a lot of these heroic acts throughout the war.

After we retreated back across the Moselle River, one third of the regiment was gone. This was mostly due to the gas shortage which gave the Krauts time to regroup and push us back across the river. Other doughboys, who were left behind on the east side of the Moselle, squirmed in their foxholes, hoping to get some help. Some others got back one or two days later. The doughboys looked sluggish, stressed out and starving, wandering about looking for food and water.

After we got more gas, we stormed the forts surrounding Metz and took them one by one. There were trenches there from World War I that we used to move to the front lines and back. Fort Driant was the largest of the forts. It was almost a mile long and went deep inside and connected to other forts around Metz. The battle of Metz lasted from November 9, 1944, until December 8, 1944. It was a tough nut to crack and a sad experience to live through.

One night while all this was going on, I was on a hill looking east down by the river bank. One company of riflemen relieved another. They walked up this hill inside the World War I trenches. I was on guard duty and it was around two in the morning. I heard voices that sounded English. The sound of foot soldiers and canteen cups approached. At first they were at a distance of about a hundred yards and then the sounds kept coming closer. Then about 50 feet away, I could see figure movement. The next thing I heard was a sergeant yelling out to me not to shoot, explaining that it was one of the line company coming from the lines below. I said, "okay, come through," so they passed by me.

After a few minutes, here came Lt. Schaech. He had been asleep inside the bunker and must have woken up after hearing the sound of voices. "Hey Lauria," he said in an arrogant voice, "How come you didn't halt the men?" I said I hadn't because I could tell they were American doughboys. He then said, "God damn it, Lauria, I should have you court-martialed." Those were his famous words. The next morning he gave me another lecture. It was the same old story over and over. I tried to tell him I knew they were American GIs, but you could forget about it. I was never trigger-happy. Lt. Schaech always went by the book. This incident happened before he lost his command as executive officer.

Earlier, I mentioned how Lt. Schaech told me he would court-martial me if I did not get my ass upstairs when Steve and I ran down and left our post during all the shelling. See, what goes around comes around. A few months after that incident, the lieutenant was up in an attic using it as an observation post with Steve when all of a sudden, the house was shelled with mortar and artillery. The lieutenant made a fast dash for the trap door leading to the floor below. With all the excitement of being scared, he slipped and fell down the ladder, breaking his leg on the landing.

It was okay for the lieutenant to leave his post but it wasn't right for me to do the same thing? Well anyway, Lt. Schaech went to a field hospital and never came back to the company. This kept him out of the rest of the war. For all I know, he may have been sent home. We still had six months of fighting to go. I never was too fond of him, but God bless him anyway.

I get a kick out of war pictures you see on TV where a plain GI would tell a sergeant or officer off. That's all crap. I know for a fact that when you were given an order, it was carried out, no ifs, ands or buts about it. There was only one lieutenant I didn't get along with, but I always obeyed his commands. I never talked back to him. Everything was "yes sir," "no sir." I never let on that I disliked him. Most of the boys in our company would badmouth him when we were alone but that was as far as it would go.

The reason I didn't care much for him was that whenever he had a bone to pick with somebody in our company he would always use the same old saying, "I'll have you court-martialed." I heard him say this to at least five or six other men in the company. Well, that's enough of him for now, thank God.

River Crossings

As a young boy, I enjoyed going to Howard Beach. I rented a rowboat and went fishing or crabbing. I remember rowing across the canal and enjoying a day out. I took my younger brothers with me. When I was overseas, I longed to be with them again, spending time in a rowboat and fishing or crabbing. But now after the war, I dread the idea of getting in a rowboat because we had to do so many river crossings during the war. I tried to go out a few times but the fear of going near the water gave me the shakes. I would see the GIs that never made it across the river. My Fifth Division had to cross over 26 rivers. Some of them were small, while others gave us a great deal of trouble. The Fifth Division's slogan was "We will" and we did.

The worst part about crossing rivers was that the enemy was always on the other side waiting for us. The Germans usually waited until we got halfway across the river and then they opened fire and shot the daylights out of us. There were times when we failed on the first attempt to make it across the river, but there were always second attempts. Try, try, try again until you succeed.

The strong current gave us another problem to worry about while crossing rivers. Sometimes we lost as many as two-thirds of our rowboats from capsizing. The loss of life took a big toll on our riflemen. The men in the rowboats paddled with all of their might while saying as many "Hail Marys" as they could muster. All of the men experienced intense fear while crossing the river, but we tried not to show it. Machine-gun fire would whiz by our heads and alongside the rowboat. I often peed my pants, sweated like crazy and of course asked God to see me over to the far side of the river.

Once we made it across the river, our problems were far from over. If we were not hit from arsenal, we may still be hit with our own mortars or artillery shells. Our objective was to advance as far in as possible and get a good foothold inland and hold until more reinforcements came ashore onto our newly won area. Riflemen fell nearby and called for medics. We were not suppose to cry or yell for help because the enemy could pick that up and shoot in our direction. If I had gotten hit, I am sure I would have called for help. It is just human nature. I was very lucky in combat. I never got hit, but was so close to being killed countless times. Shells fell around me and missed me only by a couple feet and machine-gun fire skimmed about a foot over my head. War is hell of the worst kind.

Mansion

I remember a terrible and shocking thing that occurred. On December 3 at three or four in the afternoon, the outfit had just captured some real estate. The executive officer had us stop in an area which looked like a German camp. It was the barracks of St. Avold, about 30 miles east of Metz. It had a few large buildings made of brick. They were four stories high and built really strong. Next to the building was one house that looked like a large mansion. It had two floors with an attic and was really big. I believe it was used to house the German officers and the brick buildings housed all the regular German soldiers for training. The drill field was about a 150 yards wide and 200 yards long.

After we dismounted our vehicle, we were told we were going to stay in the area for the night. Cannon Company was to occupy the large brick building. All the radio and wire men went up to the top floor to pick out a bunker but there weren't any mattresses on any of the beds. Tredanari said, "Come on, Lauria, I saw some doughboys taking mattresses out of the mansion one building away." We ran down the steps as fast as we could, hoping there would be some left for us.

We got to the mansion and entered the building. Some men were coming out with their mattresses as we went inside. We asked a colonel if we could have a mattress. He said, "Go ahead, boys, and help yourself." Then he noticed red diamond patch on our shoulder. He said, "Oh, I see you boys are from the Fifth Division." We said, "Yes, sir." He told us at one time he was a member of the Fifth Division but was transferred over to an anti-aircraft battalion. He asked us about an officer he knew. We answered that we knew most all of them in the 11th Infantry Division. As we ended the conversation, we told him that if we met up with some of the officers we would mention his name. Tredanari and I picked up our mattresses, thanked him, and left. Tredanari said, "Gee, he seemed to be a good officer."

We finally got upstairs and fixed our bed. This was the first time I would sleep on a mattress in over six months. This was too good to be true. Well, not for long. At the entrance of the building, one of the Cannon Company doughboys stood guard and another was posted around the area. I was not on guard duty that night and was happy that I would get a good night's sleep on a mattress. I went to bed and within one second I was fast asleep. I probably slept for about an hour and was in a really deep sleep.

The next thing I knew, there was an explosion so loud that it stopped my hearing. I jumped out of my bunker. My head felt like a vice had squeezed it. There was a lot of commotion going on in the stairwell. I could barely hear the top sergeant yelling out in the hallway, "Everybody out, everybody out!" He repeated it for some time. Everyone in my room called to one another to see if we were all awake since we were half-asleep, stunned and senseless. It was about 2300 hours. The top sergeant kept yelling for everyone to get out of the building.

It was a time bomb. After hearing that, I grabbed my shoes and jacket and started for the stairwell. Everyone seemed to be going down at the same time, falling on each other. I was glad that I slept with my pants and shirt on. I got to the ground floor and out the door. By the doorway was our guard, stunned and senseless. I looked at him and asked, "What happened?" He didn't answer. His eardrums were probably damaged by the blast. I looked at his carbine and the muzzle was bent like a hook. Everyone was out of the building within half a minute. The sergeant went up again to make sure everyone was out.

Outside, the area was full of smoke and dust. Since it was dark, it was even more difficult to see. Some of the engineers were doing their best to help the doughboys alongside our building. I saw one big mess. The mansion had been set up with high-explosive charges. When they went off, they demolished the mansion. It looked like a pancake. The only thing intact was the attic roof, which was lying on the debris. It was difficult to help get anyone out alive. The sergeant gave orders for all of us to clear the area around the building and go into the middle of the field. In the meantime, the engineers arrived with heavy equipment to try and rescue whoever might still be alive. They searched all night long with a floodlight and were careful not to hurt anyone. Some of us tried to help but the engineers told us to stay back and leave the job to them. They had experience at this sort of work. They thanked us for trying to help and then said, "You guys need some rest, so keep it for the fighting that's coming up."

Some of the guys waited around and others went back to the field and tried to get some sleep. Inside the mansion, there was a voice coming from the basement that all night long cried for help. Some of the engineers kept telling him, "Hold on, we are trying to get you out." The next morning, we were up really early. It was not light enough to see much yet but we went over to the mansion. The GI was still trapped inside the mansion and his voice was getting a little faint.

We stood there for awhile and walked alongside the road looking at all the dead doughboys who had been taken out of the rubble. The rubble was rough broken brick and masonry with every kind of fragment you could think of. A few radio men and I walked around the area alongside the road. The GI bodies were lined up side-by-side along the road. I started to count the bodies. I got as far as 50 when Tredanari said, "Lauria, that's the colonel lying there." He was one of the poor souls. We stayed by his body for a few minutes and said a prayer.

We continued to walk along the road and I counted 140 dead. God knows how many more they would find. They were all from the anti-aircraft battalion. God, that was a great loss. I believe it wiped out that company.

The 11th Infantry moved out within an hour. It was back to war again. Well, we were on our way to the front, which was only a mile away. The whole night was horrible. We didn't get the sleep I was looking forward to. That was a waste. But I am not complaining and I thank God that the bomb did not go off while Tredanari and I were in the mansion getting our mattresses.

The night was filled with lots of commotion and after we left it probably continued. The Krauts had done such an exceptional job on planting the time bomb that the engineers couldn't detect it. This kind of thing went on often. They kill you, you kill them. It was easy for the defense to plant booby traps and mines. They had the upper hand because they could set the traps. We were exposed to all the dangers when we tried to take new ground.

Some of the men who were sleeping in the attic of the mansion did survive the bomb blast. I talked to a few of them during the night of the blast. When the blast went off, it blew the top of the roof up and it came down like a hat on your head. There was a lot of debris that landed and pinned them down but it was not too distressing. They were able to creep and crawl slowly out of the debris. They were a few of the lucky ones. They were stunned and confused, but with the help of the engineers they were alive. Thank God for it.

IX
Battle of the Bulge

City of Saarlautern

Somewhere in Germany, near the city of Saarlautern, I woke up with a bad pain in my side. I was afraid it might be my appendix. I went to sick call around six in the morning. It was a must to go on sick call early in the morning. I went to see my top Sergeant Jones. I reported to him and told him my problem and that it was serious. He told me to load up on a truck with some other doughboys and the driver took us to an old station on the outskirts of Saarlautern.

This city was still under heavy artillery and mortar fire by the Germans, which came down around us as the truck went by. With a lot of luck, we made it safely to the aid station. I for sure thought the shelling would get me before the doctor would. The hospital was a large building which they had made into an aid station. I was interviewed by a medic and then called over by a doctor who was a major.

Right off the bat, I knew he was going to give me a hard time. The doctor was a real nasty officer and in a sarcastic voice he said, "What's your problem?" I told him that I had a bad pain in my side. He told me to lie down on this cold table which had no sheet on it. I showed him where it hurt and pressed my hand around the area where it pained me. Then the major pressed different areas of my stomach and said in a nasty way, "There is nothing wrong with you." I said that I had pain there but he didn't pay attention to me.

He called over the medic and told him to take me down to the basement. I didn't know what was going on. I asked the medic what he was up to. The medic replied that the major said to give me an enema. I overheard the doctor tell the medic, "another gold brick." I was so insulted by the doctor's remarks, I felt like busting him in the mouth.

I could not argue with an officer so I followed the medic down to the basement. I was hurt and insulted by his remarks, and I began to get angry. I told the medic to give me the enema bag. He gave it to me and I tossed it across the basement floor. Then I told the medic he should give the God damn enema to the major. I expressed my feelings to the medic and said how pissed I was at the doctor. The insult made it sound as if I wanted to get out of combat. The medic understood and I told the medic that the major could have given me a little more attention. The medic then gave me a few aspirin and felt sorry for me. I stormed out of the building and then looked for a ride back to my outfit. I had been in combat for nearly seven months and I had to hear that kind of crap from that asshole doctor.

When I got back to my unit, I told my top sergeant what had happened. He shook his head and said, "If the pain doesn't go away tomorrow I'll take you myself." The pain stayed around for a few days, and I continued to take the aspirin the medic gave me. The city of Saarlautern gave me the spooks. Death was all around it. It was dangerous getting in and out of the city to get to the field hospital. It was better to live with the pain than to get killed by the shells trying to get to the aid station.

After getting back from the field hospital in Saarlautern, I was back with my outfit, which was not too far from Saarlautern. After a few days back in the front lines, the weather started to get cold. Most of the boys were not prepared for winter. The weather had changed so fast. Also, we were so preoccupied with fighting the war we hadn't realized that winter was creeping up on us. Luckily, we did have long johns, a few extra shirts, a GI sweater, and a field jacket. I kept an extra shirt in the wire truck along with my coat, which I never wore. The days were getting shorter and colder. There was no snow yet but in a week or two we would get some. Fighting was at its lowest at the time with just lots of artillery and mortar fire going on. I was up at the O.P. with one officer to look around. When we arrived and looked around, there wasn't much to see. No enemy movement. This kind of activity went on for a few days.

Then we got some bad news. We read reports in our newspaper that a big push was starting up north, somewhere around Bastogne, which was about one hundred miles north of our sector. The Germans had broken into the American lines and advanced a hundred miles inland. This was called the Bulge. Every day it got even more intense but our division kept fighting as hard as ever.

General Patton was summoned to appear at the high command somewhere in France. He met with General Ike and General Bradley and some other high generals who were in General Ike's command. The problem was getting to the point of annoyance. The situation did not look good. So at the meeting, they asked General Patton what he thought of the matter and what he would do to stop the advancing Germans. General Patton said that he would take three of his divisions and head up north to stop the German advance. He said he would have the situation under control in 48 hours.

The Germans started this big push on December 16 and they were doing a great job of pushing one of our armies back, with great losses for the American troops. Many towns were captured in their advance and thousands of our troops were slaughtered or captured. So when General Patton said this to his general and indicated what he had in mind, all of the general's staff laughed at him and said that he was exaggerating but he just laughed it off.

He was given the go-ahead to use the three divisions so he got the ball rolling. The Fourth Armored Division, the Fifth Infantry Division and the 80th Infantry Division were given orders to pack up some time around midnight and go almost one hundred miles north on December 20, 1944. Our orders were to head north and protect Luxembourg City, which was the southern face of the Bulge. Our division was going to be replaced by the 95th Division in the Saarlautern area.

Under tight security the doughboys loaded onto all types of vehicles: trucks, tanks and whatever would move. We left in total darkness and each vehicle followed another in about 30-yard intervals. All the drivers could see was the taillight of the truck in front of them, which made the trip a difficult thing to do in total blackness. What made it even more difficult was that the road was so narrow that the vehicles just barely made it through.

I got drowsy as the night went on. I found myself waking up whenever the convoy came to a stop, which happened throughout the night. When the convoy stopped, it was standard procedure for the driver or our sergeant to come to the back of our vehicle and check each truck. The others would do likewise and check to see how the crew was doing.

Our rifles were always by our sides for fear of an ambush by the Germans who might have been hiding in the woods. This trip was a nightmare—going a hundred miles in the black of night. The weather was getting colder and the rain was turning into snow. I was happy that Turner, our truck driver, had put the canvas cover over our truck a few days back. It kept us warmer.

Eventually it was almost dawn and we could see. We were in the black forest which was in the Ardennes, Luxembourg. More snow accumulated as dawn approached. We heard shells exploding in the distance. I was getting nervous. I was sure the boys in my wire truck felt the same. I never asked anybody in my outfit how they felt. We just knew it. We would be facing the enemy soon and whatever fate it would bring to our division.

It did not take us long to realize how bad this war zone was. Looking around the area, we could see the massacre the Germans had inflicted on our troops. There were many GIs dead all over the area. I couldn't believe my eyes. It gave me a sick feeling in the pit of my stomach. There were all kinds of destroyed vehicles, including tanks, trucks, and ambulances, as well as artillery and anti-tank guns and cannons of all sizes. You don't mind the equipment being destroyed, but when you see your own doughboys lying dead, it hurt.

It looked like the whole division had been destroyed. They were lying there in the hundreds. Many were shot and killed after they were taken prisoner. It seemed as if the Germans lined the prisoners up in rows of ten or more and just shot them down. Many of them had no shoes or clothing. They had stripped the doughboys and probably used the clothing for their own use. There was destruction and horror everywhere we looked. We arrived at our destination and all of us were in shock. I was thinking and talking to some of the crew. Had this happened in the summer months, the stench from all the dead bodies would have been more than enough to endure.

The Fourth Infantry Division was getting the hell kicked out of it. I couldn't believe that our army had taken such a bad beating. We happened to arrive just in the nick of time. A lot of infantry men were trying to escape. Many were fatigued and had gone through a lot of stress. The GIs were happy to see our outfit come to their rescue. As they continued on to the rear, they thanked us for getting there.

Luxembourg: The House

Sometime during the Battle of the Bulge, after we arrived in Luxembourg on the twentieth of December, we stopped on the outskirts of the town. I remember trying to find a place to stay for the night. It was so cold and I couldn't get warm. It must have been about eight degrees and windy. Between the cold and the wind chill, I would say it felt like ten below zero. Sergeant Turner went door to door asking homeowners if they could put up the wire section for the night.

At one house, a middle-age woman came to the door and Turner asked her if she could put us up in one of her rooms. Anything would be better than being out in the cold. She told the sergeant that it was okay, but all she could give us was a closed-in porch with no heat. We decided that it would have to do. Our crew went out to the truck to get extra blankets and anything else we would need to keep warm. We were happy that at least we would have a roof over our heads.

The hayloft.

When we were outside by the truck, my friend Steve asked Turner if it was okay to give the lady some of our K-rations. We had a good week's supply. We offered the lady the rations but she didn't want to take them. She was afraid that she would take food out of our mouths. After we pled with her, she finally accepted the large box. We could tell they didn't have any food in the house. She thanked all of us and went into her own room, crying. After all, they were not German but Luxembourgian. The Krauts captured them earlier in the war and probably took their food. The next morning we were up and at it before six. We thanked the lady and she said, "God bless you and keep you safe." We left for the front lines, which were only about a mile away. Being that far away was like being home.

I remember spending one night in a hayloft. I never will forget that night. It was so cold. There must have been around 50 mice that kept crawling through the hay so it was especially hard to fall asleep. I was afraid that the mice would bite me.

The next day we were up before dawn to join the battle that the Krauts were almost finished in. I realized that Hitler was slowly going to see his maker, the Devil. He may have been the son of Satan himself. He had destroyed so much in this world. I place Stalin in the position next to him. Over fifty million humans died because of the actions of these two devils. I lost many of my friends and countrymen in this war and all of my life it will remain with me.

Luxembourg

It was almost Christmas time and we were fighting for our lives in the Ardennes Forest of Luxembourg. For me, it was a sad time of year. I know there were a good many GIs in the same boat. Those in the rear had it a thousand times better than those of us who stayed in the front lines throughout the winter campaign. There was nothing beautiful about Christmas Day. I spent it in deep snow in zero-degree weather. The wind cut into my skin. All the men that were around me felt the same. The ground was frozen and you couldn't dig a hole to get yourself safe from the weather or enemy fire. To make matters worse, we had to put up with the nebelwerfers, weapons designed to deliver chemical weapons.

We were exhausted and hungry and lacked sleep. I remember Christmas Day. We were supposed to have had hot turkey but by the time it got to our company it was frozen stiff. I had to use my army knife to break it apart. I used my knife like an ice pick. Some of my crew had a fire going. I remember putting my turkey leg on the fire for a few minutes just to get it soft enough to bite into it. I was so hungry. I didn't want to wait any longer. As we ate, the sound of nebelwerfers came at us. The noise they made was unbelievable. Every few minutes they fired the nebelwerfers at our lines. The enemy wasn't but a few hundred yards from our position.

We had no protection from foxholes because the ground was too frozen to dig. All we could do to protect ourselves from the weather was to put up a few haversacks or shelter halves to stop the wind from hitting us. All they were was a half of a tent. Two of us would get a half each and make a tent with them. In this case, it was too cold to try to make one. We might only be in one place for an hour or less, not knowing if we would advance or get pushed back by the enemy. The best thing to do, if the snow was soft, was to shovel a mound of snow and sit down in it and hope for the best. In this situation, you were better off getting hit with a shell than freezing to death.

These two pictures of fellow soldiers, whose names are now forgotten, were taken on Christmas Day. I sent the picture with the four men to Olga and wrote on the back, "My home on Christmas Day. It is really a sweet one, don't you think?" The soldiers in the top photograph are eating their Christmas turkey.

Me (left) and Joe Barber somewhere in Luxembourg in December 1944.

Frostbite was taking its toll. We called it trench foot. When it was night, we wished for daybreak. At night, the temperature would plunge. It was like a horrible dream. Our best bet was to bunk up in groups of two or three men and share the blankets and hope for daybreak to come again. This kind of warfare went on for a good month or more in January and February. I can't understand how I survived these kinds of conditions, fighting for every foot of ground!

I remember one day our unit had just occupied a town and the executive officer looking for a house to set up a command post. We entered this one house where a few ladies lived. The lieutenant told the lady of the house that she had to get out. She started to yell and cry, "You can't do this to me, this is my house." The lieutenant told her to get out. She persisted and would not go. Lt. Rocky told some of our radio crew to pick her up and put her out by force.

In the meantime, machine-gun fire and mortars were exploding in the streets. I was told to get my D.R.-8 and run a line down to the end of the street, which may have been about two hundred yards long. It ended at a wooded area across the side of the street. I really didn't know

The drunk GI from G Company.

where the captain, Tredanari and Pilip were so I decided to make a left turn on this street while the mortar and machine guns fired all around me.

As I walked with my reel of wire, I saw a GI who was at the outpost about 75 yards from me; he had a 30-caliber machine gun aimed straight at me. He yelled out to me, "Take another step and I will fill you full of holes!" I looked at him and my hair stood up from fright. I yelled to tell him that I was one of his boys. He told me not to make a move. Again, I let him know that I was with the Cannon Company running a line to the O.P. He didn't listen. Yet again, I told him, "Don't shoot. I am one of your boys." He let out of few rounds of fire but for the other side of the road. Luckily, I wasn't far from the corner building. He was yelling like a cowboy as he fired his weapon. I made a desperate dash around the corner and ran back to the command post.

Seeing Lt. Rocky, I told him what was going on down the next street. He asked me where the GI was located. I said I believe he was from G Company. The executive officer got on the phone and called up to battalion and told them that there was a drunk on a machine gun threatening

Joking around somewhere in Luxembourg in December 1944. I am the second from right.

everyone in sight, friend or foe. Well, within five minutes he was taken off his machine gun. I resumed my mission after that interruption. I guess the GI had one too many drinks of calvados or cognac.

Well, I did go forward and locate the O.P. This area was busy with German mortar and small-arms activity. The enemy was effective in doing damage to some battalion and line companies but after I arrived at the O.P., the observation men located the mortar and some of the machine-gun nests. I called back to command post for a fire mission knocking out all German resistance. The line company moved up and occupied the newly won German real estate. We stayed put until the next morning and then moved forward.

This was a day-in-day out job for me. The rest of the crew and the riflemen loved us. We were always there to help them out of a jam. They in turn protected the radio and wire crew, so we needed each other. Going back to the GI who tried to shoot me down, I still see him in the back of my mind after so many years. I was happy that he didn't get into any trouble with his company commander. Then again, I don't know what happened to him. Lots of crazy things happened in that crazy war.

X

Sauer River Crossings

Haller Luxembourg

We began to approach the city of Luxembourg. The Fifth Division was ordered to hold and defend the city at all costs in our counterattack of the Battle of the Bulge. The division established a line of defense around the city. The fighting was getting fierce. We held it as ordered. A day or so went by and we were given orders to go on the offense and attack the city of Haller on December 27, 1944. The weather had turned really bad. The freezing cold and snow with strong winds started to play a big part in this campaign.

Steve Maleevich and I had to run a telephone line to the outskirts of Haller. Our regiment prepared to attack. We set out to walk with our equipment from the company guns to the new O.P., which was some eight or nine hundred yards forward.

Everything in this area was wide open. There were wooded areas on both sides of us. The snow was so white that my silhouette could be seen a long distance and that alone made me worry. The field was at least over five or six hundred yards wide. It was dark and freezing cold. For every step Steve and I took, the snow cracked loudly under our feet. We worried about the noise of the snow crackling, which carried a long distance. The frozen snow also made it difficult because every step we took we sunk a good foot down into the snow. So we moved really slowly.

The wind was so strong that it was howling passed my ears. This made the mission worse. I only had on long johns, O.D. (olive drab), a sweater, a field jacket and a woolen cap under my helmet. My shoes were wet and frozen and my face and hands were cold and numb. Steve was no better off than I was. The distance from the Howitzers to the outpost seemed to be getting longer.

Steve talked to me as we went. I tried to answer him but my mouth was frozen. I couldn't get the words out of my mouth to answer him. We were both trying to walk in a straight line but both of us kept sinking in the snow. The footsteps were making a hell of a racket. The enemy started to fire on our position. Machine guns and mortar rained in. Steve told me to move a little faster. I complied.

The ground was frozen. Where would I take cover if the shells landed close to us? The ground was like cobblestones. At that moment, a shell passed over our heads. It landed and exploded about 50 yards from Steve and me. The shrapnel flew really close to both of us. Ten more shells came over our heads heading for the Cannon Company position and blew up about two hundred yards short of the guns. Had we walked faster we may have gotten hit with that barrage of shells. So we moved on.

We never knew where the safest place really was. Steve and I stumbled on some doughboys who were killed during the previous day's fighting. The only good thing about winter was that the stench from dead bodies didn't create a disgusting odor like they did in the summer. We now went at a faster pace and I couldn't wait to get into one of the small houses. One of them had our O.P. located inside. We were almost there and a few shells came in very close to us.

My hands were frozen stiff from carrying the D.R.-8. reel with the wire plus my rifle. A guard was at the door and he yelled out, "Halt, who is there?" Steve and I gave him the password and he gave us the answer and then told both of us to come forward. We entered the house and set up our telephone. Steve called the guns to get ready for a fire mission. In the meantime, I tried to get my hands warm. My fingers and feet were like dead sticks of wood. I was afraid to touch them, fearing they would break off.

Steve told our executive officer, Lt. Roggenstein, that he had the guns on the line. Lt. Roggenstein told Steve to tell the number one gun to stand by for a fire mission. The executive officer wanted to get a pinpoint target for the next morning or perhaps anytime during the night. After a few rounds, we located the target. Then all we had to do was wait for daybreak. Then again we may have to use our guns during the night. There is always a possibility of a counterattack.

By now, most of the riflemen, who were inside the house, were asleep. Some were on guard around the area. The radio and wire crew took turns guarding the little house, two hours on and four hours off. By the time it was our turn to go on guard again, it would be daybreak, so each one of us just had one turn on guard duty. Bogart, our switchboard operator, took a turn at guard duty. He was always at the command post to our rear.

The only one who got more sleep was our executive officer. I or someone at the C.P. would wake him up if there was an emergency or crisis. If there was no harassing enemy action, we would sit inside the house and tell stories of back home or sometimes play cards just to pass the time. Others would tell of an event that took place during the day, like a tank or jeep that had been hit with mortar fire and who the wounded were.

I wanted to get some sleep before my turn on guard duty. I wanted to forget how cold it was so the best thing to do was get some sleep. This area was isolated, with just a few houses. I was sure Krauts would surprise us during the night and give us some greeting with artillery and mortar shells. It was an easy target for the Germans for the simple reason that there were just a few little houses and I would have bet my bottom dollar it had already been pinpointed. The enemy always concentrated on such targets.

It was late December; the temperature outside was eight degrees and the windchill was ten below zero or more. It was cold inside the house. The only good thing about being inside was that we were out of the wind. We couldn't make a fire. Making a fire indoors would be suicidal. The Krauts would spot the smoke and send our boys a good 50 rounds or more. This would be the same as executing ourselves with high explosive shells.

I would usually bunk up with Sergeant Turner or Steve. If you slept alone, you put one blanket on you and one under you because the ground was so cold. Two blankets couldn't keep me warm. So our best bet was to sleep together with two blankets on the floor and two on top of us. I kept my clothes and shoes on. If you took your shoes off, it was difficult to put them back on because they would be frozen stiff. In case of an emergency, you wouldn't be able to get them back on in a hurry.

I would have to go on guard duty before long. So Steve and I decided to get some sleep. I fell asleep in a few minutes. Instead of being awakened for guard duty, we had to get up to repair our telephone line somewhere around the observation post or back at the guns. We each took a reel of wire tape, wire cutters, and our rifles, and set out to repair the line.

God, it was cold. It must have been way below zero and the windchill made it much colder. Every step of the way went at least a foot into the snow. It was about three in the morning. Each step we took made a crackling sound. Oh my God, it was cold. As I walked, I told Steve that I wished I had a cup of coffee. I meant a canteen cup. My eyes were heavy and I could have used some more sleep.

Steve held the wire in his hand and we checked it along the way. This walk had to be about nine hundred yards. As we advanced toward the guns, it was slow progress. About every 20 feet or so we would stop and tap into the wire. We were about three hundred yards closer to the guns but we still hadn't made a repair.

As we walked further along, we saw a shell hole. Sure enough, a mortar shell had hit the telephone line. We stopped to repair it and afterward used the field phone to test the line. Steve only received the O.P. but didn't get an answer from the guns. We then told the O.P. we had made one repair but still had to make contact with the Cannon Company at the command post. We started out again to repair the break. We walked and looked down at the wire for any damage. I could see the outline of the cannon about three hundred yards away. We still hadn't found the break. We moved on, holding the telephone line and approaching the outline of our guns.

Steve told me he saw another shell hole up ahead. It was about 20 yards up from where we were. A mortar shell had gotten a direct hit on the telephone line. We stopped and repaired it. I put the field phone clips on the line and Steve got both ends to answer. There were a few doughboys dead just a few feet from the shell hole. We couldn't see them from where we spotted the shell hole because they were down just about a foot drop. I made the sign of the cross and moved on.

I was happy that we had repaired the line and said to Steve, "I hope someone there has some coffee." It would get light in about an hour. I walked and yelled out to the few guards near the cannon to see if anyone had coffee. One of them said the water was hot. As we got there, I pulled out my canteen cup and a package of Nescafé coffee powder. We told the gun crew when we called awhile back that we were on our way there. Oh boy, it was cold, but the coffee hit the spot.

We started to talk about the GIs that were dead in the field of snow. It was sad. We also told them what a job it was to repair the wire with frozen fingers and feet. Steve and I were cold all over. Most of the gun crew was still asleep. They would sleep as long as possible. Sleep was scarce. After a while, Steve said, "Let's get going, Lauria." So he started out a few minutes before me. I called out to him to wait up but he wanted to get back as soon as possible.

So now we headed back to the O.P., which was up front. It was not quite light yet but you could see more in the fields. Steve was about 50 yards in front of me. In the darkness, we didn't walk too far from each other but in the daylight we felt safer and took a chance walking apart. We had both been on the run since the day before. I don't know how we got the strength to do what we did without much sleep in such cold weather. Everything on me was frozen. God help us. I walked at a faster pace, trying to catch up with Steve.

My hands and feet were frozen stiff. My hands and fingers were going to fall off. It was getting light. I heard the tanks that were parked by the little house where our O.P. was located. The tanks were making such a racket. I tried to walk a little faster but the snow held me back. I was getting a little worried because of the racket from the tanks. Tanks always made me nervous because they attracted the Krauts. My instincts told me to go faster. I didn't want to get caught in an artillery barrage.

I ran as hard as my legs could go. The snow held me back. I had about 50 yards to go in order to get inside the house. Here came the shells over my head. Just in front of me was an American Sherman tank blocking the doorway. The shrill and warble of the incoming shells told me they were going to land and explode too damn close to me. I ran faster but the shells were exploding all around me. The only hope I had was to get under the tank, so I made one desperate dive and threw my rifle and D.R.-8 to the ground. I tried to protect myself by getting under the tank.

Shells blew up all around the small area. The shrill from the shells let you know when they were going to hit really close. The road around the house was cobblestone, which was twice as deadly for a shell to land on because the shrapnel will spread out more—it explodes on top of the cobblestone and not deep into the earth. When shells blew up all around me, some of the shrapnel went past my head and hit the tread on the tank that I was taking cover under. The shrapnel traveled at about seven hundred miles an hour and could cut a human in two.

I waited a few seconds and listened for another barrage but it had stopped. I had picked myself up and started for the doorway of the house when I heard laughter coming from the tank. A GI was laughing his head off at me, saying, "What's the matter, are you afraid of a few shells exploding around you?" I got so pissed off at him for saying such a stupid thing. I got angry and grabbed my field jacket somewhere chest high and pulled on the jacket, saying, "Listen here, smart ass. Does this jacket look like

Top: **Diving under the tank.** *Bottom*: **"What's the matter, are you afraid of a few shells exploding around you?"**

it has four inches of armor plating?" The GI inside the tank looked at me, surprised and embarrassed by my remarks, and ducked down into the tank turret.

My adrenaline ran overtime. I forgot how cold my fingers and feet were. The artillery had ceased. I ran into the house and slammed the door so hard that Lt. Roggenstein looked and said, "What's the matter, Lauria?" I explained to the lieutenant what had happened outside. He listened and then said, "Good boy, Lauria, that's telling him," and laughed. Boy, was I pissed at that tank man for saying what he said. It was not called for. I had been in combat for over seven months and had to hear this wise guy tell me I was afraid. I never abandoned a mission and whenever I was given an order I did my job. Sure, I admit being afraid, but when you use common sense instead of stupidity you may live longer.

After a few minutes, I calmed down. Executive officer Roggenstein told all of us inside the room that we were going to attack the new area called Haller. We were southwest of Haller. The second regiment and 10th regiment of our division prepared to cross over the Sauer River near Diekirch.

Crossing the Sauer River in Luxembourg

At this time of year, the ground started to thaw out and the heavy rains made the field a sea of mud. Some of the company tried to get one of the ammunition trucks out of the mud. One truck was used to tow out another with its front. The mud covered the four wheels of the vehicles all the way up to the hub. The ammunition was so heavy it weighed down the truck and made things even worse. The doughboys had to take half the shells off the vehicle to lighten the load. This went on for a good hour or more.

Darkness set in. I was called to the command post along with Pfc. Campbell. We were told to get a 300 radio and go with one of our jeep drivers. He would take us about a half mile down the road to a small area with about 15 houses. There we had orders to set up the radio and wait for calls coming from the O.P., which happened to be on the hill. Communication was very poor.

We got into one of the small houses that was all boarded up. It looked like every house in the area had the same damage. All the houses

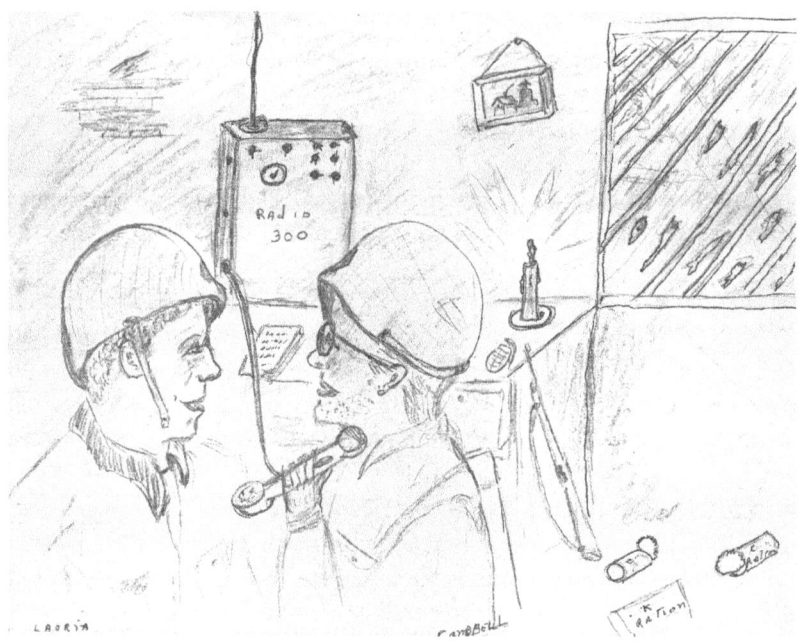

Near the Sauer River, Campbell and I were left alone in a shelled-out house.

were blown with artillery shells. The jeep driver left us there and hauled ass back to the command post. It was now so dark we couldn't see each other's face. We tried to get whoever was on the radio at the observation post up ahead.

We called and made contact with the O.P. For a few minutes, they told us to just stand by and wait. Our radio was on and Campbell and I kept on waiting. No one was in this sector. It was so scary and spooky. This went on throughout the night. Germans were on our frequency. I tried to talk to them all night long but all they were doing was trying to jam our radio. Here we were, two young kids no more than 19 years of age out alone with only our rifles and a few hand grenades. We were in a grave situation. We were both thinking, "What if the Germans come back into this town?" We would be cooked geese. We happened to have a candle and we lit it. We also tried to stuff up cracks in the boards and in the window. A few hours went by and both of us got very cold. There was no place to warm up. All we had was our one blanket. We wrapped it around us and kept trying to make contact with the command post or O.P., whoever would respond first.

Every 15 to 20 minutes we called in but to no avail. At the same time, we tried to avoid speaking too loudly into the radio. We were both scared and hoped someone would answer. We decided to take turns calling and taking small naps. If anything happened, we could wake one another up, shoot our way out and make a run for it. We had orders to stay there until relieved. Hours went by so slowly. I tried hard to keep my eyes open but I couldn't stay awake. Every once in awhile as I fell off to sleep, I jumped, almost falling off the box I used for a chair.

Oh God, how I wished daylight would come. Every so often I looked at my army watch to see what time it was. Time was standing still. Well, after a few more hours of this horrible nightmare, it finally got close to daybreak. I looked out the cracks in the window and saw nothing. Another half hour went by and we began hearing tanks in the area. Not close yet, but the sound was loud and getting near to the house that Campbell and I were occupying. Oh God, I told Campbell, I hope they are ours. It was too dark to tell.

They were in the backyard. We didn't make a move. Campbell put out the candle and I stayed put. Again, I peeked between the boards where the window once was. Oh, thank God! All I saw was one tank blocking the window with a big American star on the side. I told Campbell they were our guys and said, "Let's go out and say hello."

We went out and greeted the tank crew. They asked what we were doing here in no man's land. We both said, "Thank God you guys are here." They both scratched their heads and said, "You guys are out a half mile from the main lines." Now we started to call on our radio much louder. After about a half hour, we finally got command post. We were told to hold on and they would have the jeep pick us up. I don't know whose crazy idea it was but I was happy it was over with. By the way, there had been a lot of artillery and mortar shelling going on over our heads about every five minutes for the whole night.

This all took place after the division helped stop the Battle of the Bulge. We were one of many other outfits that had stopped the Krauts in their tracks. Sometimes when we advanced we would see many GIs dead and half covered with dirt. At times they were left with two arms or legs showing and sometimes with just the head showing. The Germans would do this to stop the smell and the spread of germs. Many dead GIs would have their rosary beads in their hands from praying while they died. It was a sad sight to witness. This happened many times during the summer.

There are times when I think back to the Battle of the Bulge, when there was so much snow on the ground. It was as cold as you could imagine. When the Germans overran the American lines, there were so many GIs lying dead on the ground. Their bodies were frozen stiff. Once we started to push the Germans back across the Sauer River, the quartermaster outfit would come by and dig out all the dead foot soldiers. They would be put on a truck and brought to the rear to a nearby cemetery and put to rest. I never did witness them being buried because my unit would be fighting too far ahead, but we all knew what was taking place. It was a sad thing and in my mind I'd think, "I could be next." This was one thought that was always with me.

Hoscheid, Luxembourg

Today I happened to drive to the newsstand about a mile from my house. When I got outside at the newsstand, I realized how cold it was outside. The snow had fallen for three hours and there must have been about two and a half inches on the ground. The wind was blowing around 25 miles an hour and the temperature was around six degrees. It was January of 2000 and my memory took me back to January of 1945 somewhere in the Ardennes. It was bitter cold that year. In my warm, comfortable bed, 55 years later, I started to think about the weather outside during the Battle of the Bulge in the Ardennes. I wondered how in God's name I survived that awful weather.

It must have been below eight degrees and the wind was blowing around 25 miles an hour. The snow was a foot high. Not only did we have to fight the harsh weather conditions, but the enemy as well. Our eating habits were not the best. We ate on the run, most of the time cold cans of food, and were glad we carried them along.

We fought for every inch of ground in a town called Hoscheid around the twenty-first of January. The G Company continued to advance in the face of 120-millimeter mortar and machine-gun fire. The enemy was in a defensive position. The G Company was getting strong enemy fire from Hoscheid. One of our self-propelled 75-millimeter guns knocked out a supporting tank destroyer. Our Cannon Company was in support of the Second Battalion and also the 19th Field Artillery. We inflicted heavy casualties on the enemy.

After getting into the town of Hoscheid, we realized how much damage and how many casualties the enemy had suffered. Dead Germans littered the ground. I was on my telephone directing the fire mission shells that landed and exploded all over the area. I didn't know what was worse, the bitter cold or the enemy. Our O.P. had just got a few close calls. The ground was frozen and lots of shells exploded. Shrapnel flew in every direction.

It was so cold. I had trouble moving my mouth to speak when I had to call a fire mission. We saw some enemy movement about two hundred yards away. It was hard to see because of all the snow. We lay on the hard, frozen ground. The snow was too soft for good protection. I prayed for the good Lord to give us some help. I hoped that the Germans wouldn't burst mortar shells on us. Riflemen were all around us and they fired their M1s into the enemy. It made a difference to me if I was occupying a small house or operating out in the open.

If we did advance into a new position, the cannons moved up closer. I dug in deep for protection. The riflemen tried to protect the radio wiring section as much as possible. If we were pinned down, I called for a fire mission. Every so often, the Germans would counterattack. The counterattack from the Germans brought great losses on both sides. Most of the time, the Germans would counterattack a second and a third time. This was the type of warfare they practiced.

The Germans almost succeeded in overrunning our position but the riflemen repulsed the attack. I remember when we advanced in an open field, working our hardest to capture a town. We had to pull back and I was told to call a fire mission. My lieutenant told me to hurry and get the guns to fire as soon as possible. All the riflemen did a backward march but they shot at the Germans as they approached our lines. The guns were ready and they were told to fire the rounds. I received another order and the second time the shells were on target. I repeated the fire missions to the command post to fire at will.

At this moment, the Germans hit our riflemen with everything they had. The Germans concentrated on hitting this area. Again, my lieutenant told me to tell the gun crew to keep firing at the same target, which was a good one. Germans were hit with our shells. The lieutenant and I withdrew and tried to get back to the next safest place. After about 20 minutes, the riflemen regrouped and started to go forward again. I saw many Germans had been hit with our Cannon Company guns. We advanced past the line of withdrawal and the riflemen marched forward. The lieutenant told me to change the reading to a new one because the Germans were fleeing the area. I had to give the guns new elevations and the executive officer took care of that at the command post.

After advancing, we secured the town. We tried to hold on to it to prevent a German counterattack. The riflemen held their positions through the night. We took turns on guard duty. Two hours on, two hours off. While I was on guard, I thought about this war and how it was a living hell. You fight all day and still have to stay awake most of the night. When I did get ready to get some sleep, it was on hard rocky ground or a cold wet surface. You couldn't be picky. I did this so many times it became a habit. Most of the time, when you got to lie down, you fell asleep within the blink of an eye. When I was awake, or on guard duty, I thought of home and felt sorry for myself. I wondered about my mother and father and hoped they would never get a telegram with the news of my death. The thought was always in my mind throughout combat. Flashbacks of combat still haunt me.

Second Sauer River Crossing

In early February, we were poised to cross the Sauer River again and enter Germany. Our observation and piper cub planes had observed near the site crossing and found that the enemy had a pillbox defense, with tanks interspersed between pillboxes. So the division had a good idea of what to expect.

The 11th Infantry was to hold the high ground. Cannon battery artillery and Cannon Company opened fire on the far side of the river, hitting the enemy. This shook the earth along the whole division front. The company moved down to the riverbank to attempt a crossing with rubber boats that we had captured from the enemy. The river was swollen due to the melting snow which was running off into the water. This time of year made an offensive launch difficult, even aside from the mines infesting the water.

The division drew heavy fire from the enemy fortification on the far side of the river, which knocked out the boats before our troops could get across. Shortly after 0800 hours, we made a second attempt to cross the river. All but two boats had sunk since the currents of the swollen river were so strong. This offensive was going to be a disaster. On the far side, the enemy had control of the high ground and this gave them an advantage.

Our division front, E Company, pulled back to reorganize and try again. In the meantime, the observation post called back to the command post for the guns to shoot the far side of the river and give our rifle company as much artillery fire as possible. The Second Regiment had a hard time making a successful crossing with all the artillery battery and each regiment raining down shells on every shoreline defense. The First Battalion marched down to the river with every gun firing.

I remember there were eight men left on the far shore, who had landed a few days before and gone undercover until we had a good bridgehead established. The eight men were exhausted but lucky to be alive. They had stayed hidden in a wooded area and had some good information on enemy troop movement and equipment. Along the whole division front, the sky lit up. I will remember this as long as I live. After the third attempt, our doughboys succeeded in securing the beachhead and gained a foothold about two thousands yards deep and eight hundred yards wide.

In 18 hours, our engineer built one treadway bridge, two class 40 bailey bridges, two assault boats and two footbridges, despite incoming artillery and mortar fire. This all happened after our troops had forced the crossing at the Sauer River. Our losses were heavy but the enemy losses were three times as much. Captain Smith, Tredanari and Pilip were across on the far side. Within an hour of the first landing, we at O.P became relayed until we were all over on the far side. Now that we were on the other side, our morale improved. We were now fighting the Germans on their homeland.

I and most of the observation crew went over much earlier than many of the other troops. We had to get there for a better view. The Cannon Company has to get close to the enemy because its range is not as long as that of the 105th Artillery. Our guns were more for mortar or directed fire on tanks. Regiment orders were always to have Cannon Company as close as possible. The radio and wire crew must always be with the riflemen in close support.

There were many near misses with enemy fire. The house we occupied had a good share of enemy shells explode around it but no sure hits. We at the O.P. gave the gun crew many fire missions. Whenever I gave an order to fire our guns, I always added to myself, "Hit the S.O.B." I know that whenever the gun crew got this many fire missions they were exhausted. By the time the fire mission was over, they had stock piled about 40 shells to each gun. They used the shells as well as the cannons. The prime mover truck driver helped carry the ammunition to each gun and throughout the battle had to avoid getting shelled as well. So the gun crew had a lot on their hands.

The shells were in a shell case which had to be opened. The shells also might need to be adjusted for a different time fuse or require smoke or armor piercing. At the time, the cannoneers had to take out a charge or two depending on distance and elevation. The shells weighed about 40 pounds apiece. The cannoneers were unsung heroes. When the guns fired, it gave the doughboys a better chance of survival. So my hat goes off to them.

After we crossed the Sauer into Germany, the Krauts were in full retreat back to their homeland across the Rhine. We were getting there fast and my outfit made the river crossing of the Rhine at Oppenheim and then again in Frankfurt. Then we experienced the nightmare of crossing over the bridge at the Main River in Frankfurt.

XI
Bitburg

I'm sorry to say that we did make many night attacks. I remember one night we attacked the city of Bitburg, Germany. As we attacked this city, it was burning down. Night fighting was difficult. Steve and I went forward with the company of riflemen and it was slow going. We couldn't see much because the city was burning and there was smoke all around. The flames from the buildings gave off some light, but it was hard to see where we were going. Our eyes burned from so much smoke but we tried to forge ahead. The smoke made inhaling and exhaling difficult. Mortar and machine-gun fire exploded close by and we tried to take cover in some of the burnt-out houses. Steve and I stayed close to the riflemen but they were as lost as we were.

Steve called back to command post to let the executive officer know we were advancing, but ever so slowly. Our lieutenant was missing for the time being, but we would find him sooner or later. As soon as we were told to hold the line, he would pop up. He would call out and someone would tell him where they had seen us. All he had to do was ask if anybody had seen the radio and wire men.

We couldn't give out a fire mission because we couldn't really see much of the enemy. They were in front of us and yet we were helpless. Our riflemen kept shooting into the enemy lines. As long as they did so, the less chance they would shoot back at us. We advanced another hundred yards or so and then the officer in charge told us to hold up and dig in until morning. It was about two bells.

The attack started earlier that night, about 2100 hours. A good five hours had gone by. The rifle company formed a line of defense and held it until morning. Steve and I called back to command post and let the officer know we would be getting some sleep. During the night, somebody would be on the radio just in case we called in. All of us up front were so exhausted and drained. We couldn't wait to get some sleep but Steve and I would have to call command post and let them know if everything was okay throughout the night.

The Krauts did not let up. They hit our area with mortar and machine-gun fire. It continued until daybreak. The morning came fast and we would attack again soon. The town was still burning and at daybreak we could see how many of our men were killed and wounded. The medics were taking care of the wounded. Some of the GIs were carried out, others walked to the rear, and others were helped by other GIs. The attack started up again and we advanced much faster in the daylight. The advance went on for a day or so. Finally, the city fell into our hands and another nightmare was over, but it had taken many of our riflemen to their graves and wounded many other GIs.

We happened to have a first lieutenant officer, Lt. Glickman, in the Cannon Company who was sent up to the lines in the advance through Bitburg. He was told to support a rifle company and he and one of our company radio men went along. As his radio observation advanced into Bitburg, they were pinned down by German mortar and machine-gun fire. The men in the rifle company were not his but sometimes when a rifle company was short an officer they would use another officer to take charge. As they went forward, the Germans opened up and took down some of the riflemen and the lieutenant. He was seriously wounded. Our radio men happened to be a good ten yards in the rear of him and didn't get hit.

As he lay there seriously wounded, he realized what had happened and the radio men called back to command post for a fire mission. Getting the call, we lost no time in calling the guns for a fire mission. Just before the attack started, the lieutenant called and asked for an adjustment and so we got a better target. The fix on a target allowed the advance to start while the lieutenant lay there seriously wounded along side the other wounded and dead.

I called back to Steve and told him that the shells were on their way. The shells were about 100 yards over and 50 yards to the right. I didn't have to tell our executive officer because he heard me repeat the order as I received it from the O.P. This way no time was lost in saying it twice and we had a jump on the fire mission.

In the meantime, the German unit was repelled and the forward riflemen advanced to a new position. The medics were taking care of the lieutenant and all the other men who were hit. Some were walking cases, others had to be taken away in stretchers. As I gave out the fire mission, the officer was taken to a field hospital and learned that he would be handicapped for the rest of his life. He had been a replacement officer in my company for a few months.

The reason I brought up this story is because President Reagan took a trip to Germany and visited Bitburg about 20 years ago. President Reagan made a stop at the cemetery to put a floral piece on the gravesite for some of the Allies and Germans who had died in World War II. There was a misunderstanding, though. Many thought he was going to put the floral piece on the grave of the SS troops, which was not true. He went there to lay a wreath on the graves of American soldiers who had lost their lives in the war and for some of the Germans who also lost their lives. There were mixed feelings about Reagan going to the cemetery. I didn't like it either, but I guess he had to keep good relations with Germany.

After Lt. Glickman was wounded my company never heard from him again. We really didn't know how badly he had been hit. During the controversy over Reagan's visit to Bitburg, I picked up a newspaper that had Lt. Glickman's name in it with an article. He bad-mouthed Reagan, explaining he was badly wounded there and that what Reagan had done was wrong. He also said he would like to stick his medal up Reagan's rear end, which I don't believe was right for him to say. I guess he had a good reason to say those things. If it had been me, maybe I would have felt the same. So I have to look at the two points of view. Then again, many of the GIs in my unit made the supreme sacrifice so others could have the good fortune to live their lives in freedom.

I sacrificed three hundred days in combat. Some days were worse than others. It was hell. I never said that my country owed me a living, but I also don't believe I owe my country a living.

When we entered Bitburg, the air force had done a great job of pounding the city for some months. Most of the buildings were destroyed. It gave me a depressed feeling to see all this around me. As Steve and I went forward with the riflemen, they shot at anything that would move in the houses. It was really mopping up whatever was still around. By this time, most of the Krauts had cleared the area. The majority of the city was secure. Steve and I walked along one street where most of the buildings were down to the ground and one or two were still intact and in fair condition. We approached one building that happened to be a musical instrument store. It contained all types of instruments. I remember taking about ten harmonicas. In those days, the Germans made the

My buddy Steve Maleevich in Bitburg.

marine band harmonica, which was a popular instrument in the U.S. As a kid, I learned to play one. I took about ten of them and put them in every pocket I had. When there was a lull in the fighting, I had time to ship them home to my brothers who were much younger then I and had not been drafted. The sad part about this was that when I got home to the States nobody seemed to have the instruments. I remember finding just one harmonica that was in good condition left in the house. Nobody knew where they went. I'm sure at that time they were worth three or four hundred dollars. Some of them were a foot long while others had double reeds or even triple reeds. Some of the instruments in the instrument store were too big to carry around so I just left them behind. I was surprised that the fire hadn't destroyed everything inside the store. This town left me with bad memories.

As we went along with the riflemen, I came close to getting hit again by some small-arms mortar fire. We were dodging in and out of some buildings that were not altogether down to the ground and provided some cover. The lieutenant, Steve and I went along with the foot soldiers. If we were pinned down, we would call a fire mission. As we went forward, we stumbled with every foot of ground we gained. The small-arms fire never stopped. If I have said this once, I've said it a million times. It was hell. The sweat was running down my face and I wet myself. I couldn't think. My mind was blank. I thought I was too scared and too excited to communicate my command to the command post, yet I managed to do so.

Shells exploded all around the area. GIs were trying to take cover

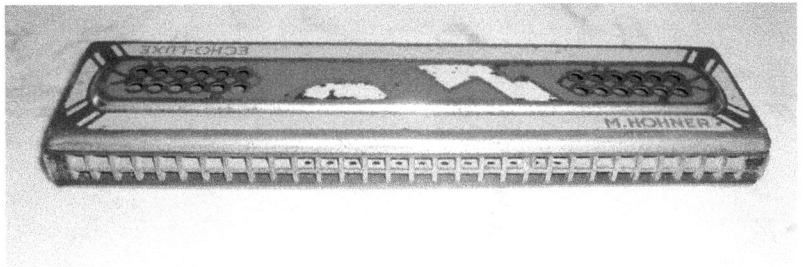

The one remaining harmonica from the group that I shipped home.

but most of the buildings were destroyed so there was no place to hide. The walls on the houses were almost to the ground. There were no doorways to get into. Our only hope was to pray and hope the enemy mortars were located so we could knock them out with our cannons. I moved ahead.

This mopping up went on for some time. I don't really remember. The city had fallen into our hands after a few days of fighting. After the capture of Bitburg, we moved on to a new area of war. This was just another hard-won city that the Fifth Division had taken. When Bitburg made the news after all those years, I thought, "How about that. I was one of the GIs who helped capture the city."

Ring

My buddy Tredanari gave me a German ring. He had gotten it off some other doughboy along the way. He came back to command post and had a few S or what we called loot. I said, "Hey Len, how about giving me that ring you have?" He stopped for a moment and said no. I asked again and again. Finally, he gave me the ring. He also had a German fountain pen. It was an old 14-karat gold pen with a bladder which held the ink. I guess in those days it was worth something. Anyway, he gave me both of them.

After a few days, I asked him where he had gotten the ring. I figured he looted it from inside a German house or had taken it from a German P.O.W. He said it had been given to him by a GI when he was up front at the O.P. He then told me it came off a dead German's finger. I said, "What do you mean? He pulled it off his finger?" He said, "No, the GI had to cut it off the dead man's finger." After he told me that, I felt a

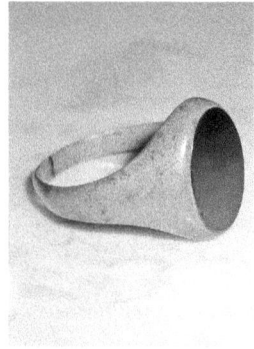
The ring.

little squeamish about the ring. After seeing so many dead Germans, I should have been used to this going on. Instead I got in a cold sweat about the matter and wanted to give it back. Well, I didn't. I put it inside the truck seat until the end of the war. I took it home with me but I always got that squeamish feeling about it. Sometimes I would see it somewhere in my box of collectibles and say a prayer for whoever it belonged to. Sometimes I wanted to bury the ring although it is nice looking. It is gold and has an onyx stone.

I was never a GI who went looking in the clothes of dead bodies for loot. I saw many other GIs search for loot in the pockets of dead Krauts but I never had the stomachs for it. I saw more dead bodies than I want to talk about but never tried touching them. I have eaten within a few feet of them but never put my hand on them. I was always squeamish when it came to that sort of thing.

Dud

On one particular day, we were in position in a field about two hundred yards wide with trees lining it on all sides. The cannons were spaced out as always, about 10 to 15 yards apart. For a while, there was a lull in the fighting. Some of the boys got into their foxholes and tried to get some needed sleep, which was hard to get. Some men would just mope around and others would pick up a *Stars and Stripes* newspaper and read.

After about one hour, some mortar and artillery shells came in on us. Steve and I were up at the O.P. and got word that the guns were getting attacked. The shells exploded in the gun section 10 and 20 yards apart. Up at the O.P., we looked to see where the enemy guns were firing from. Another barrage of shells came in. They were exploding close but we couldn't seem to locate the guns. About 15 more rounds came in after this. All of a sudden, one of the men in the gun section started to scream. It was a sharp, shrill scream and he repeated it for some time. This made most of the cannon crew look out of their foxholes to see what was going on. This GI kept on screaming for a few more minutes. He started to explain what happened and pointed to his foxhole but wouldn't go near it.

After he got his head together, he explained what had happened. He told everyone that a shell had landed between his legs while he was in a squat position. The shell had gone through his newspaper and his legs. The shell penetrated about a foot into the ground. As soon as this happened, he jumped out of the foxhole and started to scream.

Later, he realized how lucky he was to be alive. The shell was a dud. He thanked whoever made the shell. It must have been a prisoner of war. There were many of them helping in the war effort. They would do whatever they could to sabotage the enemy.

When I got back to the gun crew, I talked to the GI. He explained that when the shell hit, he was so scared that he jumped out of the foxhole in one leap as far as his legs would spring him, realizing the shell could be a time fuse. He thanked God for having a prisoner of war sabotage the shell. The Germans never could find out how many shells didn't explode. I guess this helped save a lot of our boys' lives.

Bazooka

Somewhere in Germany, during a different battle, my company occupied a small house and used it as a command post. Our guns were in the rear of the house. I finished laying my telephone line to the guns and went back into the house to hook up the line to the field phone. The forward observation was about four hundred yards in front. I didn't go forward that day since the area was flat and we could use radio signals.

Corporal Hicks was on his radio and went forward with Lt. Jayson. There were rifles moving up the line of fire. The Germans spotted our riflemen. The Krauts opened fire on them while I was busy hooking up the telephone line to my field phone. Artillery and machine-gun fire showered down on the house. This incident happened so fast, and all hell broke loose. Our boys were in a vulnerable position out in the open field. The German tiger tanks emerged out of the surrounding woods and blasted away at our riflemen who were trapped in the open field. The German riflemen were behind their tanks, shooting at everything in sight.

Some of the shells hit the house that we used as the O.P. I heard Corporal Hicks call me for a fire mission. I rang up the guns and requested a fire mission. I told the executive officer that Lt. Jayson wanted a fire mission. All guns responded and were ready but the enemy was too close to our guns. Hicks yelled over the radio for some backup. He told me to tell Lt. Roggenstein to call regiment and get some air support.

Some doughboys fired their bazookas and 60-millimeter mortar shells, which did some good. In the meantime, most of the men stayed pinned down and hoped the Cannon Company would give them some support. Hicks was still on the radio as Lt. Jayson gave Hicks the location to fire our cannon. Hicks seemed to be in a panic while he was on the radio.

I felt hopeless. I hoped the planes would get there soon. Some of the riflemen had put out their panels so the airmen would know exactly where they were. Both Lt. Jayson and Hicks were almost hit with machine-gun fire. Hicks told me to tell Lt. Rock to keep the guns firing. The guns were doing the best they could so we continued with the fire missions. After about half a minute, I did not receive Hicks. I was afraid that some of our shells might have rained down on our troops. I didn't know what to do.

I told Lt. Rock and he told me to have the guns at a different elevation and fire at will. We changed the coordinates again. I didn't know if Hicks or Lt. Jayson were hit up at the O.P. There was some radio silence. We waited for some word. Then, we heard the P-47s as they struck the German tiger tanks. Six or seven P-47 planes dove down on the tanks that were about one hundred yards from the riflemen. Some of the German riflemen jumped off the tanks and made a run for the woods.

The tiger tanks were under attack but were still moving. As the Krauts ran for cover in the wooded area, one of our bazooka men got hit. He was mortally wounded when a tiger tank came straight for him. Hicks, Lt. Jayson, and a platoon were useless against a German tank. Hicks crawled towards the wounded GI and picked up his bazooka, which had one remaining shell. He yelled to Lt. Jayson to load it as he aimed it at the tiger tank. Hicks had only fired a bazooka one time before in Ireland. It was something we all had to do for training. With all the excitement, Hicks aimed the bazooka at the tiger tank while Lt. Jayson loaded it. Hicks was shaking but tried hard to hold it steady. Lt. Jayson hooked the wire to the bazooka and tapped Hicks on the shoulder to let him know that the bazooka was ready to fire. Hicks aimed again and fired the rocket. The rocket doesn't go as fast as a bullet and it wobbles as it flies. The shell hit the target and exploded, hitting the tank's tread.

The Germans ran outside of their ruined tank and ran for cover in the nearby woods. Our P-47s were superior weapons and shot at the Germans. They took out two tanks while Corporal Hicks and Jayson hit the third tank. Hicks got on the radio and told me to hold our fire. I didn't get the full story until we saw Hicks and Jayson back at the command post a few hours later. In the meantime, darkness set in and the rifle company took the high ground.

We had no idea how dangerous it was when all hell broke loose. The tanks fired at the few houses we occupied. The walls of the houses were over a foot thick, so the machine-gun bullets just bounced off the walls. The cannon fired very close to the house and missed it by a couple of feet. Luckily, the Germans didn't fire at the roof, which was the weakest part of the house. Since I kept busy with the telephone and was distracted by the airplanes overhead, I didn't have time to realize the magnitude of danger that we were in.

It is hard to control your fear when you know the enemy is so close. The house had no windows on the side where the enemy was and we had no idea how dangerous it had gotten. There was a lot of commotion when the enemy fired at us. The planes shot back at the tanks and our guns did likewise. We were in the command post giving fire missions, while trying to hug the floor. When I wasn't taking care of my duty, I prayed to God to help keep me alive. I thanked God that I didn't have to go out and repair a telephone line that day since it was all radio talk. Thank God the P-47s repelled the attack and Corporal Hicks and Jayson used the bazooka to take the tank out of action.

Three other German tanks did get away, though. Had it not been for the air support, the Krauts would have overrun our position and then gone in for the kill at the command post. Then it would have been up to the Cannon Company to fire at the tanks point-blank. If it came to this, I am sure that the Cannon Company would not have been able to take the Krauts out of action.

Thank God the Germans stopped about three hundred yards away from us. There wasn't time to retreat. Everything happened so fast. Actually, we never used the word retreat in the U.S. Army. We always said withdrawal.

When Lt. Jayson and Corporal Hicks returned to the command post, they told us how the battle had taken place. Jayson reported to the regiment. The army awarded Hicks a Silver Star for his action and Jayson received the Bronze Star. War is a strange thing because you might be trained to be a radio forward observer and find yourself having to use a bazooka to kill the enemy. It takes some training and a lot of guts or, shall I say, courage.

The credit must go to the riflemen, the army air force, and the Cannon Company, who stayed at their guns and completed the fire mission, and of course to the men inside the command post, who did a magnificent job of staying at their posts when the tension was unbearable. We all worked as a team in the command post. Because we were in a small house with nothing but fields around us, the Germans had a good idea that there was a command post in there.

XII

Lieg

March 16, 1945

The Fifth Division advanced near the Rhine River in Germany. We fought our way deeper and deeper into enemy towns. I remember we were not too far from the Rhine River when our unit came to a halt. We were told to dismount and get ready to dig in. This area had steep hills on both sides. I didn't know if we were going to stay and dig in for a while or move out as soon as our riflemen went past our position. I waited for Sergeant Turner to give me orders. I wanted to know if I should run a line from the guns to the C.P. and up to the O.P., wherever that was going to be.

In the meantime, I was with Pfc. Owen Stanley, who was our company recorder. We did not have any idea what was going on. I got off the wire truck. Sergeant Turner told us to wait for him to come back from the guns. Our executive officer, Lt. Roggenstein, was there so Sergeant Turner went to get things going.

The wire truck was parked about ten feet away from a slit trench which was dug by the Germans, who must have occupied the trench earlier. Pfc. Stanley asked me if I wanted to bunk together, if we were going to stay there for the night. He asked me several times. Instead, I made a new foxhole for myself and Sergeant Turner. Again, Stanley asked me to bunk with him. I told him to make the slit trench a little larger and the lieutenant would stay with him. Funny thing is, sometimes we make the slit trenches side by side, but this time I made my hole a foot away, just below Stanley's.

Sergeant Turner came back and I asked him what we were going to do. He said that he had no news yet but that I should dig a trench for both of us, which I had started to do. Turner told me he would be back from command post once he found out if we were staying or pulling out.

He went down to the cannons which were about three hundred yards behind my foxhole. I finished the hole in no time flat because I used the large shovel which we carried on our wire truck. Then I ran one wire down to the guns and one back to where Pfc. Stanley and Lt. Stanley were making their hole larger.

This was going to be our command post. Pfc. Owen Stanley was our company recorder and Lt. Stanley Anderson was a new officer in our company. He would learn what goes on in the command post during a fire mission. Lt. Anderson had been in charge of a whole rifle company in our regiment but was coming down with a nervous breakdown so he would stay with the Cannon Company.

After I finished with my hole, I had put back the shovel that belonged to our wire unit but forgot to put my helmet back on my head and left it near the rear right wheel of the truck, about ten feet away from my slit trench. A few minutes went by and a shell came in on a hill on our right side and exploded some hundred yards away. I looked and gave it little attention. A few more minutes went by and another shell exploded on the opposite side, on a hill which was about 400 feet high and about 150 yards away. This time I did give it a little thought. I told Pfc. Stanley and the lieutenant that I believed a German observation was trying to zero in on our position.

I looked to my left and there was a dead German about eight feet away. Some of our riflemen were looking in his jacket and pants pockets for any loot. This was never my cup of tea. After some time went by, a few more riflemen came over and also rolled him over and looked in his pockets. At this rate, he would be rolled all the way up the hill.

I spoke to Stanley again, saying I didn't like what I saw with the shells that had exploded on both sides of us. Then Pfc. Stanley started to dig deeper. Five more minutes went by and about ten shells exploded all around our area, too close for comfort. We had all taken cover in our holes. Just about the time they exploded, Sergeant Turner jumped into my slit trench and held me down. After this happened, I wanted to get up and get my helmet, which was about ten feet away. The sergeant told me to stay put and I did.

The shells landed all around our area and a few landed by the guns that were about two hundred yards to our rear. I wanted to retrieve my helmet in the worst way. When the helmet was on my head, I felt like I was under a pill box. The sergeant called out to Pfc. Stanley and the lieutenant to see if they were okay. They answered, "yes." A few more minutes went by and again I wanted my helmet. In the meantime, Lt. Bass was on the other side of our wire truck, but we didn't know this yet. Sarge got up to see if everything was okay. He started to walk over to retrieve my helmet and another barrage came in. This one was nearer than the last. So the sergeant jumped into the slit trench and held me down. I began to feel a little madness in this situation.

There was a lull in the air and again I made an attempt to get my helmet. I was so nervous that I couldn't hear the shells coming in. The shells were so close that they affected my hearing. The shells had to be 105 millimeter. I knew they would be on top of us. The sound was so frightening. I started to piss myself, not knowing I was doing so. I felt that the shells were going to land on top of us.

Another barrage of shells came in and this was so far the worst of them. I was digging my hands into the ground, trying to use my fingers as a shovel. The concussion was so bad it picked Sergeant Turner and me off the ground. Turner held me again and asked if I was okay. The ground trembled and shook under me and we both waited to see what would happen. Sarge asked me again if I was okay. I said, "yes," but I knew something had gone wrong around us. The sergeant picked himself up to look around to see the damage.

He called over to Lt. Anderson and Pfc. Stanley. Turner was on his knee. There was no answer. The next thing I heard was, "Oh no, oh my God, oh my God." I knew something bad must have happened. He told me not to look but I knew it was bad because there were parts of flesh and blood on me and Turner. My carbine was full of flesh and parts of bodies. Sarge called for the medics.

I looked and saw something that will remain in my brain forever. I said to Turner, "There is nothing the medics can do." There were parts of bones in my trench and clothes. When the explosion had gone off, everything had fallen down on the sergeant and me. Pfc. Stanley was hit dead on his lower torso. His arms and the upper part of him were still intact. His intestines were hanging out of the upper part of his torso. I could see his arms still moving. I couldn't believe my eyes. I had been talking to Stanley a few minutes before and now he was dead.

I was getting hysterical. I was emotional and started to cry. This thing was bad. He was like a brother to me. Lt. Stanley Anderson's body had been blown out of the slit trench by the concussion. The lieutenant's head was split in half as if a man had used an axe on him. His arms were moving and his brains were out of his skull. On the other side of the wire

This scene stayed in my nightmares my entire life.

truck, Lt. Bass's head had been cut off from the neck by another shell that had exploded just ten feet from me.

Shrapnel played a big part in this barrage. I paced back and forth. I couldn't bear to look anymore. Sergeant Turner ordered me down by the guns. I had no control of myself. Some of the gun crew rushed up to see what had happened. Some of them turned away. I remember someone putting a blanket over Lt. Anderson's and Pfc. Stanley's bodies. Turner put his arm around me and walked me down to the gun position. Two of my buddies pulled me down to a safe place to stay near the cannons, which were also hit.

Lieg

after the heavy mortar shelling Sergeant Turner had taking me down by the gun section some two hundred yard to the rear leaving the horrble sight where Lt. Bass & Lt. Stanly anderson and owen Stanly P.F.C., Sergeant Turner tells me to stay put under the lumber for safty. darkness had come ismethung. I cryed the whoel night. a nightmare that is still with me till this day 2006 They ort Knowing that some Twenty or more of our men are hurt

Turner told me to stay put under the lumber for safety.

Some 25 gun crew cannoneers were wounded. All I saw in front of me were Lt. Bass and the two Stanleys, dead in such a horrifying way. The wounded cannoneers were rushed to the aid station. I just couldn't stop crying. My nerves were getting the best of me. Some of the doughboys tried to control me.

There was an area nearby that looked like a lumber yard. When it started to get dark, Sergeant Turner told me to stay put under the lumber for safety. Some of the planks were sticking out a few feet and had good overhead protection. I stayed there the whole night and was so upset. I remember the whole night long I kept thinking of Stanley, his arms still moving on just half a body.

Sergeant Turner looked for me and wanted to know how I was doing. Morning had come and we were to move. With the lack of sleep and the nightmare hanging over me, I was in a daze, stupefied by what had happened. All night long, I mourned and grieved for the loss of my friends. I still didn't know what had happened to our gun crew. It wasn't quite light yet when Sergeant Turner came to wake me up. Little did he know that I hadn't slept a wink that night.

Sarge had to lead me around for almost a week after that horrible moment of losing my good friends. It took that long before I snapped out of it. Fighting was going on and many towns were captured but I do not remember any of them. I did start to remember when our unit was getting set to cross the Rhine River and when the crossing was made.

My mind started to clear a little but the sadness is still with me. It has been almost 60 years. One of our men lived near Pfc. Stanley in North Carolina and they always talked of seeing each other after the war if they made it safely home. After Stanley was killed, I told Pvt. Martin, "If you ever go to visit Stanley's mother and father, don't tell them how he was killed. Make it simple." So after the war, I spoke to Pfc. Martin and he told me he would only tell them he had been hit with a bullet and died, but I never did find out if he ever visited Stanley's home.

Lt. Stanley Anderson was in the 11th Infantry Regiment over four years. I knew him a short time. Lt. Bass was with the company a few years and was a wonderful guy. I remember giving him a German flag that was over 15 feet wide and 30 feet long. I didn't want it because I was superstitious. I believed it would bring bad luck so I gave it to Lt. Bass. I know he mailed it home a week after I gave it to him. I remember the large box he mailed it in.

It is really mysterious how things happen in life. Lieutenant Anderson was taken out of G Company, which he commanded, and put with the Cannon Company to give him a break from his line company. You never knew where it would be safer. To me this was God's work. This goes to show you that when your time is up, it's up. He was in the wrong place at the wrong time. Many times I wonder how we could have prevented this nightmare but no one has the answer. They say time heals all. I guess time helps, but it always come back to haunt me.

Had he stayed alive Lt. Anderson would have been a forward observation officer in our company. The few days that I knew him I tried to make him comfortable. I felt so bad for him because he was sent to us to get him away from his rifle company. He was a good officer. At the retreat back across the Moselle River, he went back and forth to retrieve his men and stayed to the very end. He, Lt. Marshall, and Lt. Genci were the lasts ones to leave for the western shore and safety. They were real heroes.

Getting back to the incident in which our company was hit so hard and so many of our men were lost, it took me about 60 years to learn of the town where this nightmare occurred. I happened to see it in a book that my son Ken had requested from the library. After receiving the regiment book, we found a small article stating what happened on that date, March 16, 1945. I learned what town it was near. It was called Lieg. Now I should have remembered this but since I was in another state of mind at the time I didn't find out until 2002. Everything else in my stories came from my mind. When the unit got a story in the book, it was cut down to a few lines and that was about the size of it.

So many times I asked myself why we had to take up position in that particular area. Maybe it could have been prevented. We were in the wrong place at the wrong time. After a few days, some of the boys came back from the hospital. They told us who was hurt more than others. It made me feel good to hear the news of their recovery.

Sergeant Forrest Turner is on the right. He saved my life three times in combat. The other man is Bill, last name unknown.

I stayed in a trance and appeared to be absent from my duty. I guess my wire sergeant understood and looked after me for a few days. That tragedy was too much for me to handle. I had lost someone really close to me. I was in a daze. The incident left me stupefied. I don't remember any of the towns we captured for a good week afterwards. I knew almost every town we went into or captured for the simple reason that I always kept a map of places we were going to attack. The company gave maps to the radio, wire men and the officer in charge.

Sergeant Turner kept an eye on me for the four or five days that I was still in shock over the loss of men in my company. The sergeant was like my older brother. He made sure I was okay. Around March 22, we made an attempt to cross the Rhine River. My mind was getting clearer but thoughts of that horrible day were still with me. They are still with me today.

After the war, I went to church and lit a few candles for all my company men who had died in combat. I did this every Memorial Day for a good 20 years or more. I don't go to church to light a candle much now, but sometimes I do. I still wake up in the middle of the night with this nightmare. I was the one who gave Stanley the foxhole so I keep blaming myself for his death. This has gone on and on since the year 1945.

Once again, it was Sergeant Turner that saved me from the same horrible fate. If I had gotten up to recover my helmet, the shrapnel from the last barrage would have killed me. Thanks a million, Sergeant Turner. I came so close to death myself. I don't understand how I survived that hell. God, I guess, and Sergeant Turner, who was the angel on my shoulder.

XIII

Oppenheim: Crossing the Rhine

Now the Fifth was ordered back into the Third Army again. We were never taken out, just loaned to one of the other armies. We headed south again. General Patton wanted badly to cross the Rhine River and so the Fifth Division got the honor of making the crossing. It was March 22, 1945. I believe our crossing of the Rhine was the Divisions' twenty-third crossing. At the banks of the Rhine River, we waited to make the crossing. All of us were sweating this river because I believe it was the widest the division had attempted to cross. I'm sure it was over nine hundred feet wide.

The top brass ordered that no one open fire on the enemy in order to maintain the element of surprise. There was talk of the navy helping us with some large landing craft. Well, on the night of the crossing, two rifle companies of the 11th Infantry and K Company went over in assault boats. The crossing was a success. The element of surprise paid off. The 11th Infantry established a bridgehead on the far side of the Rhine River. After a few hours, the bridgehead was widened and the regiment gained more ground.

This crossing was the first assault crossing of the Rhine in history. The Second Battalion was in reserve at the town of Dachau, which was on the west side of the Rhine. After establishing a bigger foothold, the engineers started to build bridges, but there were many ups and downs in keeping the bridgehead secured. Many more of our men gathered over the footbridges and soon the engineers had completed the heavy bridge so we could get our armored unit and the 4th Armored Unit over.

First our attack armored unit came over to help widen the bridgehead. Our rifle company was doing a wonderful job. I remember going over the large bridge and the convoy was jammed with all kinds of vehicles. At the foot of the bridge were two black anti–aircraft guns. We

11th Infantry crossing the Rhine River with K Company making the first wave assault.

seemed to be at a standstill and here we were about five hundred feet out and nowhere near the far end. The current was swift and there were a few low clouds hanging overhead.

One major was standing on the pontoon bridge, telling all the drivers to hurry it up. Some other wire men and I were in our wire truck when six Messerschmitt 109s came from out of nowhere, shooting at the bridge and the convoy. I believe the planes were using 20-millimeter guns to open up on us. I looked up and the 109s dropped their bombs. We jumped off the trucks, tanks and jeeps, you name it. We were stuck on this pontoon bridge which had been built in a half a day.

In the meantime, everyone was looking for cover. I was under the vehicle. The 109s missed the bridge and the bomb exploded in the Rhine River. Other planes did likewise—none of the 109s hit their target. The major kept yelling for everyone to get back in their vehicles. We got back on and hoped the Jerries would not make another pass at the bridge. It happened so fast, in just a split second.

The battalion major yelled again and told us to get going across the pontoon bridge. It seemed like everyone started to go in a big hurry. The column moved at a much faster pace. As we got to the far end of the bridge, here came the planes again. The first time they made the pass, they were in a straight line as the Rhine flowed. In other words they flew parallel to the Rhine. On their next pass, the 109s were coming in from the west, straight over the bridge, strafing the whole length of the pontoon bridge.

At this moment, my bunch of guys were just getting to the end of the bridge. Luckily, the 109s missed our crew, but I don't know if any of the other vehicles were hit with the 109s' 20-millimeter machine-gun fire. Our truck driver was doing his darndest to get away from the bridge and make it to the small town. Once we were across the bridge, it was dreadful as he steered from side to side on a road which was full of shell holes. We were about 50 feet away from a barn when another 109 started strafing the road. We all jumped out of the truck and headed for the barn. I didn't know how safe it was going to be, but that's where we went.

I made a desperate leap for the barn door and the plane came around and struck again. My body was almost completely inside the barn but one leg was still on the outside. The shell from the 20-millimeter struck the ground about five inches from my foot. Some of the rocks and dirt hit my leg and shoe. I started shaking and trembling. I realized how close I had been to getting hit with the 20-millimeter gun. I waited a few minutes to see if the Krauts were still there but they had taken off for home.

Most of the gun crews in our army shot back at the planes. Anti-aircraft claimed they shot down two 109s but I don't really know for sure if that was true. My crew jumped back in our wire truck and proceeded forward. Our rifle company had advanced a good mile. In the meantime, the Cannon Company had gone almost all the way to the front lines. We were told to jump off the trucks and get the guns set up for a possible fire mission. I set up my telephone wire to the guns and headed up forward.

The Krauts were hitting us with machine-gun and mortar fire and we could see their position so the lieutenant told Steve to get the guns on the radio. I was there and hooked up my telephone to call the guns just in case the radio went out on us. So now we had back-to-back communication. The Germans counterattacked our position. Resistance had stiffened. We were on the outskirts of a town called Wallerstädten. This was March 24. B Company was holding down a counterattack and C Company was holding back another attack on the right of us.

I was amazed to find how close our Cannon Company was to the enemy. My company would usually stay about four or five hundred yards behind the riflemen but this time they were right up with the riflemen. I realized this after the battle. Usually, I was up with the foot soldiers and didn't know what was going on behind me.

I was scared most of the time in combat but after a battle, I would start to shake. I realized how close death was. There were times after a battle that we would talk about what had gone on there. Other times, I would hide and cry until I let it all out. Being uneasy was part of being in combat. Sometimes things got so hot, I was weak in the knees. We usually hid our fears. Most of us had courage when given an order; we would carry it out even though we were always scared.

We were now in the town of Wallerstädten. As the fighting went on, the enemy started to fall back and we advanced again. I was relieved and told to go back to command post. I realized how close the cannons were to our front lines. Most of the time I was forward and didn't get to stay long with my company. I was a stone's throw away from the Krauts in every battle. It was nerve-racking. I trembled in each battle, scared to death. I shook more when it was over than when it actually went on.

There were many times when I would hide in some area to be alone, get down on my butt and let it all out. Crying to me was a comfort because there was no one to help me out of that mess. Whoever was there was in the same fix as I was. I prayed to God to see me through all this misery and begged him to give me courage to make it from one day to the next.

Prisoners of War

I remember one area where our unit had just finished taking a town. The company had taken position in a field. I started to walk to a large building that looked like a hospital but I wasn't sure. I walked a path that led to an iron gate. It was about four feet wide. The gate was held by brick pillars. One side had hinges and the gate was latched to the other. I went forward to the gate but used precaution because I had no business being there alone. When I got to the gate and swung it open, it made a squealing noise.

At that moment, I saw some prisoners of war inside. Some were American and others were British. I remember how happy they were to

Oppenheim

He kept repeating, "God bless you guys" over and over.

see me. One airman had been shot down some two years before. All of the men were yelling, "God bless you! We are free!" One of the airmen was walking with crutches. His left leg was amputated after he had been shot down over Germany. He came over to me and dropped his crutches and hung on to me, crying and kissing me. I had tears in my eyes and told him our troops had arrived and that I was a forward observation radio man. I told him to stay put and help would be on the way.

I had to leave and go forward because there was shelling going on. He said he heard the shells but they seemed like they were a few miles away. I asked him if there were any Krauts in the building. They said most of the German army had left but only the German doctor and his staff had stayed behind to care for the sick and wounded.

I didn't stay much longer. I estimated that three to four hundred of our guys were in the hospital. After a few minutes, I said my goodbyes and they said, "God bless you and may God keep you well and safe." I waved to all of them and went on to the next objective.

Until this day, I can still see the airman coming over to hug and kiss me. The prisoners never knew what was going on or how close we were to them. They never got any news about the war but they said they had a good idea our boys were getting close to them. When new prisoners were taken, I guess they would give the old ones some news on how much the Allies had advanced. It was a good feeling to see the prisoners so happy.

XIV
Main River Bridge in Frankfurt

After making the crossing over the Rhine River, my division captured some small towns and villages. We had not yet captured the city of Frankfurt because there was an obstacle standing in our way, the bridge over the Main River. This bridge would lead us into Frankfurt. The bridge was a hindrance in the drive to capture the city. The Jerries were going to hold the bridge at all costs. Fighting stiffened and it seemed like every German was armed with a panzerfaust and every other weapon in the German arsenal. The day we made the crossing I believe was the 26th or 27th of March. At this time, my head was still in a spin over the loss of Lt. Bass, Lt. Stanley and Pfc. Stanley. That horrible day had been about ten days before.

The gun crew and other members of our company who had been hit returned to our company. I had bunked with Stanley many nights and the loss of my buddy hurt me. For many years, the loss was horrifying and to this day it is a nightmare. I will feel the loss of my company men until I am gone.

Now the company was waiting near the side of the bridge. Many riflemen were going to try to fight their way to the far side, which, if I am not mistaken, was over six hundred yards away. It was not going to be a piece of cake. Only God knew what would happen. Fighting had intensified during the daylight hours. Our riflemen made an attempt to get across. Just a few men managed to get across. Resistance mounted as the day went on.

In the meantime, on the far side, enemy tanks, mortar fire, artillery and enemy machine guns kept hitting our men. The bridge was called the Remagen Bridge. The Jerries used time bombs to try to blow it up but the attempt didn't work as the Krauts wanted it to. The time device went kaput. Some of their bombs did blow up and left large gaping holes on the bridge. Some of them were over 25 feet in diameter. When our company made an attempt to cross the bridge, we discovered the holes.

In the meantime, a few riflemen made slow and sure headway. Mortar and artillery hit the bridge. I don't remember ever hearing as many mortar and artillery shells come in as I did on this day. We waited for our captain to give orders to go across but fighting was more intense than ever. Going over at this time was sure suicide. The Krauts used their 88-millimeter guns, which were deadly. The explosions were harrowing. Also, they used heavy mortar shells. The Jerries were using everything but their kitchen sinks.

So, we waited until it was almost dark. Captain Smith gave Lt. Roggenstein the word to go ahead and attempt the crossing. The enemy fired almost directly at our unit. I suppose they could see our small truck coming. I am sure it wasn't dark enough. Most of the boys on our wire truck looked for the large holes in the bridge. Sometimes, a German flare lit up the sky and would help our truck driver and the members of my crew see the gaping holes in the bridge. When we got near enough, we would all yell to Pfc. Turner, our truck driver, to be careful. Not only did we have to be careful with the gaping holes, we also got pounded by the shells and machine-gun fire coming in on us. My heart was in my mouth and so it was for everyone who was making this suicide run. The going was slow.

I wanted to jump off the truck and run across but we went along with whatever was going to happen. It was getting much darker now so the Krauts used a few flares to try to see us. This ordeal was not getting better. We progressed slowly, keeping our eyes on the road, trying to stay safe from shells landing and exploding near us. The 15 or more minutes seemed like hours. This was a nightmare's nightmare. After a while, some of the riflemen climbed the truck and went along with us. We traveled in a zigzag motion. The drive was even slower because it was not straight, and so the nightmare went on.

After what seemed like an eternity, we reached the end of the bridge. The enemy machine guns still fired at us and our riflemen jumped off the trucks and returned fire. Most of the Krauts in this sector had moved to a new position. Lt. Roggenstein located a good place for the guns. I set up my wire to each one and waited for the lieutenant to give me orders on where our O.P. would be.

Thank God the night was dark. It helped us get across the bridge much more safely. Luck and opportunity were in our favor, although the battle was far from over. The whole night was nothing but shells going and coming. Throughout the night, we constantly heard the sound of shells exploding close to us and machine-gun fire just over our heads. The captain never gave us orders to look for him or to go to the O.P. In a way, I was glad I didn't have to go. It was bad enough where we were. I didn't get a minute of sleep that night.

I don't recall if I had eaten that day. My body was so weak, I didn't have the strength. Exhaustion was the right word. Digging a foxhole felt like too much to do, but I did manage a shallow one. The continuous pounding of artillery and mortar fire was unbearable. Explosions were too close for comfort. I trembled throughout the night. Morning was coming. I could see the sky getting lighter but the pounding of artillery and mortar and machine guns still went on.

I thought back to the bridge when it was so dark. We had to go through one sector where some apartments were five or six stories high. Krauts were shooting from windows and rooftops and our riflemen were on both sides of the street shooting back at them but they managed to get out of there safely.

A few doughboys got hit by Germans on the rooftop. We were told later that some of them were civilians and others were Hitler youth and old Volkssturm. We also learned that some civilians with binoculars were directing the fire. Anyway, the night made it much easier for us to get across alive. There were Germans occupying buildings with 88s and anti-aircraft guns, which have a high velocity. They also used tanks. Each one of our regiments got across which helped mop up the enemy. There were over five hundred thousand civilians in the city of Frankfurt, but to tell you the truth I didn't see many when we occupied it. I guess most of the civilians were in hiding.

It was much more difficult getting across the Main River than the Rhine. The Krauts had stiffened their resistance, plus, there was a lot of house-to-house fighting, which everyone in my outfit disliked. The constant pounding of enemy mortar and artillery was unbearable. There was also mice-hole fighting from room to room and on rooftops. There were snipers everywhere. Riflemen were looking out for snipers in windows and on rooftops, wherever they were hiding and shooting at us. Whenever a German shot at a GI in the dark, you could see where it was coming from because of the flash from his rifle and the doughboys would fire back. Our men were on both sides of the street protecting us as we advanced through the city streets.

Mice-hole fighting was the worst type of warfare. The constant pounding of mortar and artillery fire was unbearable. The fighting hadn't let up since we crossed the Rhine River; we had a few small towns to mop up. They were tough. We had many causalities before we took Frankfurt.

Lt. Roggenstein led our company into an area which looked like a field with a road on it. I realized later that this was a runway for airplanes. We got set to dismount and the cannons set up their guns. The radio was clear and the captain was on the radio. Tredanari was going to give us a fire mission. To my surprise the radio was loud and clear. Steve received one fire mission after another. I couldn't believe that we were sending out so many shells. The tanks were 50 yards to the rear of us and the cracking explosive sound was unbearable. This went on for a few hours. It was near the break of dawn.

Steve Maleevich was on the radio with Tredanari. I was on the phone repeating the fire missions to the guns. My voice was hoarse from so much talking on the telephone. It had gone on and on. As dawn approached, I could see an object that looked like an airplane. I was talking to Steve and said, "Isn't that a plane?" Steve said yes and pointed over to where a few more were located. As the light increased, I recognized that the area we were in was an airport. I could see some fighter planes in the distance. Now I knew why there had been so much activity during the night. The Krauts were trying to hold on to this airport. Then Steve said, "Now I know why my radio sound was so clear last night." It was all open space and I guess Captain Smith and Pfc. Tredanari and Pilip were not too far away from us.

There was a battle still going on at the Frankfurt railroad station, which our doughboys were having a hard time capturing. Captain Smith told Tredanari to tell Pfc. Malcevich he would give him a new reading on a fire mission. Pfc. Maleevich repeated it to Lt. Roggenstein. The lieutenant told me to give the number one gun the new reading. Then I got the new coordinates to our guns. Each cannon had one man getting my command. They listened and each gunner took the same command but only the number one gun would execute the command when I gave the order to fire his gun.

Captain Smith had the Germans in sight. He wanted to pinpoint the new target. The captain told Tredanari to give Steve the new reading. It seemed like the new target would be about two hundred yards away from the old target. Lt. Roggenstein calculated the new number one gun and told me number one gun was ready. I repeated it to Lt. Roggenstein, then the lieutenant told me to execute the command. If the number one gun was on target, Captain Smith would have all six guns adjust to number one gun and then I would order the six guns to fire if it was a big target. Things were going well for us.

Germans were still holding out in some parts of the city. Others got out of the city and moved to new positions. It took at least three days to clear the enemy out. We suffered many losses but the Krauts suffered a lot more. As we finished mopping up inside the city, we were told to take a breather. At the end of March, after our regiment liberated the city of Frankfurt, we were given a day or two of rest.

All of the company took some time to wash up. If we had any clean clothes, we put them on. It sure felt good to be clean. Our kitchen and cook with its staff had come up to the lines and made the company a hot meal, which was a blessing. Most of the time, we had K- or C-rations to eat. This hot meal was like going out to a restaurant. We were thankful to receive it. I thanked God for sparing my life again. The good Lord was always with me. I never let him out of my mind.

After our company had eaten, Steve Maleevich and I took a walk by the railroad cars where our unit had had a hard time the other day. It was a nice day and the weather was getting warmer. Steve and I walked along the railroad tracks and looked inside each railroad car. Some of the doors were closed. Steve and I opened most of them and climbed inside. The cars had medical equipment for special purposes. We went along looking in some of them. Some of the cars had locks. Steve and I took our carbine and tried to shoot off the locks from about a hundred feet away. I missed every shot, not because of my eyesight but because my gun barrel was bent. I know because I had a marksman and expert medal so something was wrong. Boy was I happy to find out. Thank God I didn't have to use that carbine on some Kraut. The Kraut would have gotten me before I got him.

Steve and I went back to our company and told some of the men what we had done at the railroad yard. As we walked through the airfield and saw some of the fighter planes that were shot up by our shells, I climbed up on the nose of one fighter plane. Steve snapped my picture on the plane along with some other shots. Later, I sent the photos to regiment to get developed. Some of the photos were sent back but the one with me on the nice fighter wasn't there. I believe the regiment kept some of the important pictures for army use. I remember it but don't have any proof of my story. After all these years, I think about that snapshot with me on the nose of the fighter.

In the distance I could hear gunfire, but it was about a mile away. For us dogfaces, this was a long way from the front line. In the meantime we were relaxing, but I knew we would be on the move soon. A day later we received orders to move up north to the Ruhr pocket and try to capture the hundred thousand or more Krauts that were still holding out. Well, that is another story. The month of March was over and we were going into April. Little did we know, F.D.R. would pass away and the war would be over in Europe within a month and eight days.

The bridge crossing of the Main River was one of my worst nightmares. I did lots of trembling in that crossing and the explosions were unbearable. It was a hair-raising, scary experience. I would not wish it on my worst enemy. Crossing over a bridge gives the enemy an advantage. If you were pinned down, there was no way to dig a hole with your shovel. We just hoped that when the shells exploded near us we would be spared from the high velocity or acceleration. German 88 guns were what I feared the most while on the bridge. The Krauts used air bursts overhead and there was no way of protecting ourselves from the blasts. Shrapnel rained down on us and the only cover we had was if we stopped the vehicle and got under it. So the best thing to do was move on. That was better than being a standing target.

Our unit did some damage in Frankfurt but most of the damage was done by the B-17s. I don't remember if I saw one building left standing in the city of Frankfurt. Whenever the Allies knew a city had industrial and productive defense work, the allies would pound those cities to the ground. Most of Germany was in ruins. I believe the Krauts had to learn the hard way that they were not the master race. I believe the night on the bridge was one of my worst nightmares. There wasn't a moment when the shelling stopped, it went on and on through the night. Now I bring this chapter of crossing the Main River bridge to a close. Frankfurt was now ours.

Starvation

I remember that during the fighting in Germany there was starvation going on. We didn't see much of it when we were fighting because most of the civilians were hiding in deep cellars or out-of-the-way places. Now, whenever the unit stopped for a day or so and one of our other regiments would take over our area, we would be given a day of rest. Most of the time, the captain would have the mess sergeant prepare a hot meal for the company. I guess the regiment had done the same for the other three battalions.

We would get a good hot meal. It could be anything which we usually hadn't had in months, including poached eggs, mashed potatoes or maybe a pork chop if we were lucky, but this was hard to come by. We also would have real coffee, which I always had seconds of.

I remember that if we were near a town there would always be five or ten little boys or girls around us, staring at all the GIs as we ate. I felt sorry for them and so did most of the company men. We all had our mess kit full of food and, to tell you the truth, getting a hot meal was a blessing. Most of us would go up to the open table where the food was given out and ask for seconds. If it was there, you got it. Our cooks didn't stay up front with the company but they sure gave us whatever they had.

I recall that after such a meal, we would go to the garbage can that the cooks brought with them and dump whatever was left. No sooner did we do this, than the German kids would rush over to the garbage can and scoop up leftover food with their hands or with a small pan that they had with them if they were lucky.

The part which hurt me was that most of the GIs smoked and when they were finished with their cigarettes they would snuff it out in their mess kit, leftover coffee, mashed potatoes, or whatever else they didn't eat. It was all dumped into the garbage can and the little German kids would run for it like it was just cooked. They would scoop it up and eat it as they put it into their pots. So, starvation was there.

We were not allowed to give them any food because they were the enemy and we were not allowed to fraternize with the enemy. I know in my heart the young kids were not to blame for Hitler's doing. I would see this going on and always had a tear in my eye for them. This was a shame. They were born at the wrong time. Well, this was just one of the sights I experienced in combat that I couldn't believe.

The Fifth Infantry Division

The famous Fourth Armored Division in the Third Army was highly publicized. I am not trying to take anything away from the outfit but it seemed to me like they got the most credit for everything they accomplished. During my time in combat with the Fifth Division, we accomplished many outstanding deeds of valor. We were given the honor of making river crossings and yet so little was said about our success. In the three hundred days in combat, the Fifth made over 26 river crossings and not much was written about them, from Normandy beach to rivers in Austria and Czechoslovakia.

The Red Diamond Division made the crossings of the Kyll, Moselle and Rhine Rivers. The Fifth made three successive bridgeheads. On March 22, 1945, when the I and K Companies of the 11th Infantry Regiment of the Fifth were the first to cross the Rhine, not much was even said about it.

When the Fourth Armored Division crossed on our bridgehead and then dashed forward, they made the headlines, while the Fifth trailed behind them to mop up all enemy resistance. You must remember the riflemen only had their clothes to protect them where members of the armored division had four or five inches of plating protecting them.

There were many outstanding divisions in the ETO doing a magnificent job, but to my knowledge many of the outfit received little recognition. It seemed to me that just a handful of the Divisions have always gotten into the news and they were always the same outfits. Some were from the Pacific, others from Italy, and some from the Normandy invasion. I believe most people don't know that there were about a hundred different divisions in the ETO and that the lines went from the English Channel to the south of France and included British, French and Canadian soldiers. One outfit could only cover about a mile or so of ground. This helps people understand that it took more than just a handful of divisions to win this war.

I could go on about the outstanding job that so many of my outfit buddies did. One Fifth Infantry Division member received the Medal of Honor, 34 received the Distinguished Service Cross, 9 received the Legion of Merit, and 602 received Silver Stars. Members of the Fifth Infantry Division also received 2,067 Bronze Stars. I was honored with a Silver Service Star and two individual Bronze Star medals.

Ruhr Pocket

After Frankfurt, the Fifth was ordered to head up to the Ruhr pocket. The high brass believed that over two hundred thousand German

Depiction of American and German infantry men and officers.

troops could be in the area. So the Fifth Infantry was ordered to race up in that direction to try to surround the fleeing enemy.

At that time, I realized that the war was coming to a close. The Krauts were giving up much more easily and seemed to be disorganized. Lots of the Krauts were giving up in groups. Don't get me wrong, the fighting was still going on. We were losing men just as fast as before, but we were gaining more ground doing so. I believe that after the Battle of the Bulge, they ran out of everything they needed. We kept mopping up whatever was in our path. Retaliation on their part seemed to be slowing down but the war was still not over.

XV
End of the War

It was sometime around May 6 or 7, 1945. The war was coming to a close and I occupied a house with some of my radio and wire crew. The command post happened to be about two hundred yards to the rear of the house. The guns were another two hundred yards from the command post.

I decided to clean my rifle and my German P-38 pistol. I noticed that all of the boys who had been with me had left the house. I occupied myself by cleaning my guns. I sat on the front steps with the guns that I had taken apart. There were some army trucks going by with German prisoners of war in them. This type of activity went on for several hours.

After a while, the trucks stopped coming and everything got very still. On the other side of the road, there was a wooded area that ran parallel with the road. It became so quiet that you could hear a pin drop. The whole front line was doing nothing. I called the command post to see who was there and they told me everyone in the wire crew was at the command post. I guess they went there for a change of scenery. One of the men should have stayed with me, but no one did.

I looked to the right of the woods and saw some Germans making their way towards me. I thought to myself, "Oh shit! I have my guns all apart and here come some Germans!" I had to do something fast. I pushed part of my rifle and most of my P-38 into the hall and grasped the holster and put my hand on the handle to make it look like I was ready to shoot the first German that made a bad move. As they came closer to me, all of them said, "Comrade" and put their hands on their heads. They must have taken their helmets off in the woods and thrown them away.

When they were within ten feet of me, I yelled, "Halt!" and all seven of them came to a sudden stop. I spoke to them in the few German words I knew. They answered back, "Comrade." It seemed like they all spoke at the same time. They did look a bit frightened. I was so scared that I felt

Depiction of the Cannon Company throughout the ETO.

like going blind. I didn't have a gun! I told them to come with me. I made them walk in front of me with their hands on their heads. I kept my right hand on my pistol, which was only the handle in the holster. I wanted to make sure they kept thinking that I had a real gun in the holster.

I walked the prisoners over to the command post and some of my buddies in the radio section said, "What have you got there?" I explained to them what had happened. One of the GIs called Lt. Rocky and told him that I had seven prisoners. He saw them and said, "Goddamn it, Lauria, get rid of them. We have enough to do without nursing the Krauts." So the lieutenant told me to bring them back to a road that went west and tell them to keep walking. After the Germans left, I told my buddies what had happened with the gun and they couldn't believe it. I went back to the house and finished cleaning my guns.

The next day we moved to a new location somewhere near Austria. The date was May 8, 1945. There were still pockets of resistance. I remember this for a fact. The resistance was still going on for a few days. The night of the tenth of May we were told the war was over. It felt really good to hear the news. I remember sleeping in the open air and not worrying about enemy shells or gunfire. My worries seemed to be gone. That night I slept under one blanket and didn't have to dig a hole to sleep in. It was such a wonderful feeling. All of our company was jubilant.

Many of us still carried the sadness of the war experience with us. We stayed up and talked about the men we lost in the company. I remember I got on my knees and thanked the good Lord for sparing me and said the Lord's Prayer for those who did not make it through the war. Back at home some were happy, but many people dealt with grief for those they lost during the war. It was both a happy and sad time for those of us in the air force, navy, and army.

Fistfight

I want to write about having a fistfight with one of the GIs in my company. I believe the war had just ended. The Allies had announced that the war had come to an end on May 8, 1945, but my outfit was still fighting the Krauts on May 10, 1945. We were up in the Bavarian mountains and I guess they hadn't gotten the word that the war was over. At least the army told us to stop and just hold our position.

I never told anyone about this because they may have said I was making this cockamamie story up, so I just let it go and say that the war had ended on the eighth of May. At that time, resistance was negligible and disorganized but the enemy was still around. There were some smalls arm fights and we had some difficulty with the mountainous terrain. I remember seeing the movie *All Quiet on the Western Front* and I didn't want anything like that to happen to me. The last shot at the last moment of war.

Well, our unit was given orders to move to this new area. My company took over a large house which looked like it had housed troops because there were some large rooms about 30 by 40 feet. In these rooms, there were about 20 single beds.

Tredanari and I picked two end beds and were so happy the war was over. Most of us just couldn't get over how happy we were. Some of the guys were singing and playing cards. Some of the guys played a game of craps. Our happiness was just too great to explain.

Tredanari and I decided to get some eggs and potatoes and fry them for a meal. As we were getting our lunch ready, I went outside the door and saw a table, which we put between our beds. There was just enough room for the table to fit. After setting up the table, we went outside to cook our eggs and potatoes and then carried the food into the room and sat down to eat.

As Tredanari and I sat down to eat, a member of our company comes over to our table. I'm sure he most have had a glass of something to drink. His name was Osthoff. He may have been 28 or 30 years of age. He started to hit my shoulder a few times. I stopped eating and told him to stop hitting me but he didn't. This went on for a minute or two. I kept telling him to stop but he just kept it up. I told him again and again. He then said to me, "What are you going to do about it? I told him to stop or someone would get hurt. He then asked me, "You think you are tough?" I said, "No, I don't want any trouble." He then took a swing at me across my face. It just so happens I had my fork in my hand with some food on it almost in my mouth. He took another swing at me and his fist hit my fork and the fork almost went into my eye. It hit the skin between my mouth and my nose.

At that moment, I was very annoyed and angry from the pain of the fork in my skin. I tried to get myself out from between the bed and the table. Momentarily, I was stuck, but Osthoff kept swinging at me. I just had one way to get out from between the bed and the table as he was still hitting me. I put my two hands under the table and picked it up. It toppled over, spilling the food all over Tredanari. I got myself free and as Osthoff was still hitting me, I blocked a few punches and then set myself up to fight back. I now had some swinging room and I let out a right punch and hit him square on the side of his mouth and jaw.

I know I hit him hard. He fell back about eight feet on his back. I was over him and I saw his eyes were wide open and blood was coming out of his face. He was out like a light. He wasn't moving. I got scared and ran out of the room, looking for a place to hide. I can't remember where I went but I know it was a few fields away. I stayed there for a good two hours and then headed back to my room, where the fight had started.

I was still shaken. Sergeant Turner saw me and told me not to worry. Osthoff was going to be all right. He wasn't back yet from the aid station or field hospital. The sergeant told me to sit down by my bench and pretend like nothing had happened. I was too shaken up to pretend. Finally, here comes Osthoff into the room with two of the company men helping him along to his bunk. I looked and saw what I had done to him. He had a large bandage around his head, all the way under his jaw and covering most of his mouth. He kept saying, "I have a good idea who did this and I will have it out with him as soon as I get better."

Me, on right, boxing with Len Tredanari for fun somewhere near Metz. Owen Stanley is on the far left.

Sergeant Turner told me in a low voice not to answer him but he kept on talking about how he would get me. I wanted to fight him again but the sergeant told me to say nothing. Osthoff kept this up for some time. The same words kept coming from his mouth. The sergeant then said to him, "If you don't keep quiet, I will give you the same thing Lauria gave you." Sergeant had given me away but most of the boys in the room started to laugh. I am sure Osthoff knew what had happened to him.

What I think made Osthoff come after me was the fact that in Ireland and sometimes during combat, I would box with some of my buddies just to pass the time. One of my buddies knew a professional fighter by the name of Freddy Sevic and used to box with him. I knew of him before I went into the army. My buddy would show me how he would box with him. We used to box a lot in Ireland. As a matter of fact, I have a picture of me boxing with Tredanari. One day I was boxing with Osthoff and I got the best of him, but in a joking way. I never wanted to seriously fight my own buddies. Fighting the Krauts was bad enough.

This fight spoiled the ending of the war for the simple reason that I had the scare of almost killing Osthoff. It stayed on my mind for some time. The war's end was now secondary in my life. The fight made me scared. I am sure the combat time and the fistfight contributed to my

This photograph was taken in Soissons, France, while I was in the hospital.

having a nervous breakdown. After getting sick, I never did see him again. I have no hard feelings for Osthoff and if he is still alive I wish him well. If he is around, he must be about 85 years of age.

When I had the nervous breakdown, I was taken to the hospital. I lay on a stretcher for three days. When I awoke, I was told by a medic that I had lain in that stretcher for almost three days as if in a coma. I was never myself afterwards. I was put on a plane called the DC-3 and taken to a hospital in France in a town called Soissons. After a week or so, I begged the doctor to send me back to my outfit. I saw the doctor twice a day and he would talk to me and ask me all kinds of questions. I would tell him I am okay just let me go back to my outfit. After pleading with the doctor for a week, he said that I would be sorry if I ever signed myself out of the hospital because if I ever got a recurrence, I would never get compensation. He was right. In later years, I tried to get some compensation from the VA hospital but never could because I signed myself out.

After getting back to the outfit, I really was never the same. I had to rehabilitate myself. It took many years. No one but my wife understood the problem. When I returned home, my boss, Carl Peckler, the owner of American Almond Products, told me to come back to work to take my mind off the war. I have to admit that my wife was the biggest cure. We loved each other and this did help.

XVI
Home

I signed myself out of the hospital because I desperately wanted to go home. The day I got out of the hospital, I walked 25 miles to Camp Saint-Lôuis near Reims, where my unit was stationed. They were headed back to the States. I had so much loot but after I had a nervous breakdown, everything I had in my duffle bag was gone, just a few things that some of my crew had of mine were left. They gave it back to me when I returned from the hospital.

I was just happy we were headed home. We got 30 days furlough. We set sail for the U.S. on July 13, 1945, and I arrived home on July 20. When I got home, I saw a newsreel of Americans dancing in the streets of New York and all over the U.S. when the war ended in Europe, but the war wasn't over yet. After I had spent about 16 days in the States, the U.S. dropped the atomic bomb on Japan. Three days later, we dropped another bomb. At the time, I was on my way to my girlfriend Olga's house. The radio announced that Japan surrendered while I was walking to her house, still dressed in army uniform. I was on 101st Street. People ran into the street, happy and yelling that the war was over. Some of the crowd tried to pick me up to celebrate. They were all in such cheerful spirits. They applauded and danced in the streets. I put my feet on the ground and ran as fast as I could over to my girlfriend's. I ran about ten blocks. She was waiting for me by her front door. We kissed and hugged and kissed. We said to each other, "Thank God it is over." The next year, on April 28, she became my wife.

The bomb changed everything. There had been plans to send my company to Japan, but luckily the bomb put a stop to the war. We were going to train on open tanks with Howitzers mounted on them. After the war ended, nobody in the outfit was interested in doing any training. This went on for a month or so. All we wanted was to be discharged.

Finally, on October 9, 1945, I was discharged. It seemed liked a strange coincidence because I left the states on October 9, 1943. I landed in England, went to Ireland and then landed in Normandy on the ninth. I also started the journey home from France to the United States on the ninth of July. It seemed like the ninth had a big part in my life. That date also played a large part in the army. I received a Bronze Star on August 9, 1944, and another on a date that I don't remember. I am disappointed that I didn't write all of this down 50 years ago. I would have remembered almost all of the details.

Sometime after the war, I happened to be at a party with some of my friends and cousins and we had a discussion about the war. They asked if I ever killed anybody in combat. I had to stop and think for a few seconds. I replied that I had killed many Germans in combat but not with my rifle. My weapon was my radio and phone. Fighting alongside our riflemen when we were pinned down, Steve, our lieutenant, and I would take over and try to blow out whatever was holding us up, whether it was a machine-gun nest, a mortar crew or a German platoon of riflemen. We would be ready to try to knock them out with our small howitzer that used a 105 shell.

The rifle platoon was happy to see us alongside of them and we were happy that the riflemen were alongside of us. The short cannon was a good weapon. The gun crew would have that cannon in position in a few minutes. It was good for knocking out tanks, pillboxes and houses with the enemy inside of them. I said to my friends and cousins that whenever I was at the O.P. with my lieutenant and Steve and there was a fire mission, you would rest assured that someone on the front line was going to get killed. Sometimes it was a few and sometimes the numbers were in the dozens. In Chartres, in about a half hour, we killed over 250 Germans and wounded 800.

My killing was something like when a group of bombers go on a raid. The bombs drop and who gets the blame? All of us, I guess. This put me the a situation of wondering who the killer is. I had something to do with it and there is nothing I can do about it. So, I leave it up to whoever asks me that question. As for me, I trust the almighty God and his judgment on my wartime activity. In a way, I am happy that I never killed the enemy with my rifle because it has given me some peace of mind. I know whichever way I look at it I am still at fault. I always pray to God to forgive me for my acts in war. Whatever the reason, I try to resolve the right or wrong feelings by praying to the Almighty himself and I hope I am forgiven when the time comes.

I was responsible for killing enemy soldiers indirectly. I gave the orders to fire the guns and saw so many of them die because of my words. In a way, I still have a feeling of guilt. I will make peace with my maker, but I feel that I am still at fault.

The fighting was the same thing over and over but each day was a new challenge. We were exposed to gun, mortar and artillery fire whenever we set out or went forward. It was pure hell. I went that route almost every day in combat. If I went to sleep at night, I thanked God that I woke up the next morning. Nightmare or not, I was glad to be alive. If we were pinned down by the enemy, my hand always held onto my rosary beads, which were in the pocket left of my heart. It's funny how you can pray and have them come out so automatically, the Our Fathers and Hail Marys. I can't explain it, you have to experience it for yourself. It isn't like the movies when they show a man using a gun or any kind of weapon and the shooter avoids a thousand rounds of shell fire and walks away with no harm done to him. In real life, you piss your pants, you crap yourself and you shake like a leaf on a tree in high wind.

I thank God I was never hit, but I did come so close to it. I can't imagine how I survived the hell I went through. There were situations so bad I wanted to take my helmet, pull it down over my ears all the way down to my shoes and stay covered like I had a pillbox over myself. I got this feeling whenever we were in an artillery bombardment or mortar barrage or whatever else they decide to hit us with. Again, thank God I am here writing about my life in ETO combat.

Now after some 50 years or more, there are times I awake in the middle of the night and think about the hell I went through in World War II. I stay awake and wonder about the many GIs that died in combat and the thousands of airmen shot down out of the sky. I witnessed so many bomber and fighter planes burn on their way down to earth. Some were airmen with only four or five missions while others with 15 missions behind them got shot down too. You may get killed on your first mission or live to your twentieth or thirtieth. It was a matter of luck. Take Normandy, for example, there were thousands who died without making it to the beach, drowning in the ocean, while others made it 40 or 50 feet inland and didn't survive either. Others would last a week, two weeks, or a month and others were lucky enough to survive through the entire war.

Sergeant Turner passed away some 20 years ago. I used to write to him and send him a Christmas card each year. One Christmas I received a card from his wife telling me he had passed away from cancer. Sergeant

Turner saved my life three times and I thank him for it many times in my prayers. Henry Bogart also passed away from a heart attack about the same time Sergeant Turner passed away. There were a few others I would write to but they stopped writing or lost contact.

When I got sick at the end of the war from the nervous breakdown, I lost my address book. I never did find out where most of the boys lived. My biggest disappointment was that I wasn't able to see Tredanari. I had his home address and I tried to get in touch with him. His sister answered the letter but he never tried to see me. I also went to Philadelphia, his hometown, but never did find him. I guess he didn't want to communicate with me. I tried for almost five years but never did get an answer. It really hurt to know he wanted nothing to do with me.

I did write to another buddy who lived a mile or so away from Tredanari and he told me he went to see him. It was a cold reception. Something must have snapped in his head because we were so close in Ireland and in combat. He received three Bronze Stars. I just can't understand why this friendship had to come to an end.

The other buddy who went to see Tredanari at his house had gotten hit by shrapnel. One hunk of it went through his helmet and rolled around inside between the helmet lining and the shell cap and came out the other side of the helmet. He was one lucky S.O.B. Afterwards, he was dizzy for a few weeks. I know for a fact that he took the helmet home for a remembrance. I did write to him for awhile and but eventually stopped.

In closing this saga, I want to express my feelings for the men who died for their country. They were the true heroes who made the supreme sacrifice, whether they were a foot soldier, a GI in the rear or a medic who was there in a time of need. They put their lives on the line throughout the war. The forward radio and wire men were also unsung heroes at the forward observation post. They were all heroes, the airmen who flew the fighter planes and bombers and the engineers who had to build the pontoon bridges while the foot soldiers or riflemen held a bridgehead on the far side of the river so the heavy tanks and reinforcement could get over to widen the bridgehead. Yes, they were all unsung heroes of this war. God bless them all. I believe much more should have been done for us. We were the sacrifice.

I made it but do not have a reason why I survived. I thank God that I am alive. The hundreds of thousands of military graves may give you an idea of what took place in World War II. Each of these individuals gave their life for someone else's freedom. I believe that people do not

Me and my girlfriend (and future wife) Olga when I returned home.

Me and my mother and father when I returned home.

care enough for the GIs who put their lives on the line, enabling others later on to enjoy the freedom that this great country of ours offers. I wonder about the many mothers and fathers who lost children and other loved ones. I imagine how they felt and how they lived with it forever. I remember that during the three hundred days that I spent in combat my mother and father suffered many sleepless nights and worried constantly about my safety. It was a one-day-at-a-time situation. You live minute to minute with the help of God and everyone at home praying for you.

I often say to myself, "Lou, how in heaven's name did you survive?" I wonder what my mother and father went through during the time I was in combat. It baffles my mind how they were able to endure this nightmare. All I know is it took hard praying day in and day out. I am happy that my folks did not receive a letter from the War Department stating that I was killed or missing in action.

I had over three hundred days in combat and it never got any better. It affected my entire life from the day I was drafted. It changed my attitude toward life. You don't know how precious life is until you could lose it at any moment. Every day that goes by I thank God that I lived through that experience, and I appreciate life no matter how hard it gets. The experience I lived through brought me closer to God. It renewed my faith. I also mellowed out. You truly do not want to get into trouble or hurt people after living through the horror of war.

Every town or city we captured has a story, but those stories would take a lifetime to write. I'm happy to put some of this down in writing. In years to come, my grandchildren may want to know what I did in World War II.

Finally, I just want to express my feelings about my stories of World War II. After 50 years, I can still feel the hurt in me for losing so many of my friends. I pray for my fallen comrades. May God be with them forever. I will have their memories with me forever and I thank God for giving me the extra time to go on with my life. I am proud to tell you I was part of the greatest generation that ever lived. If I would have to be called again, I would answer the call as I did at the age of 18.

Afterword

After the war, Louis returned to his home in Queens, New York. He was fortunate and didn't lose any immediate family members in the war, although a few distant cousins lost their lives. He quickly proposed to Olga and they were married on April 28, 1946. To help keep his mind off the war, Louis returned to work for American Almond Products.

The war was with Louis his entire life. For many years after the war, he was very nervous. When he first got home, he had difficulty going over bridges because of his bad experiences in the war. His anxiety made it difficult for him to swallow sometimes. Boats, bridges and rivers were all hard for him to be near because of his combat experience. When he drove over bridges and the road rumbled, he was always reminded of planes strafing the bridges in Europe.

He never answered the phone because it reminded him too much of fire missions during combat, which would drive Olga crazy. He would sit next to the phone and just let it ring off the hook. She eventually got used to it and understood that it was difficult for him. After the war, Louis never flew on an airplane. The plane ride he took at the end of the war on the DC-3 was his first and last. He associated his trauma from the war with that flight and refused to fly again.

It always bothered him that the war didn't change people or society and our country continued to engage in conflicts. Louis strongly believed that if his generation hadn't stepped up and fought for democracy, Hitler would have committed genocide all over the world.

Although Louis said he would willingly serve our country again, he often felt that veterans were not given enough for their sacrifice. When he returned from the war, he disembarked at the port in New York City. His family didn't know he had arrived so he had to walk home. He tried to hitch a ride as he walked from Manhattan to Queens but nobody stopped. This hurt him deeply.

Louis risked his life and lost many friends in the war and he often felt his country neglected him. In a letter Louis wrote to an editor in 2002, he explains this sentiment:

> When I got discharged, I stayed home for about three weeks and then I went back to my old job to try to forget all the bad and sad things that I had to tough out in the war.... Yes, I received medals and citations; but believe me they did not help my hardship. Maybe I was a fool because I never collected my 52/20 money. I thought I was being a nice guy; but what a mistake that was. In 1997, I had open heart surgery and the government wasn't there for me.... This isn't fair to those of us who put our lives on the line for freedom.... What about the GIs buried on foreign soil and the ones who survived with serious wounds and cannot get any help or respect? The United States has thrown all the GIs to the wolves.

Despite his disappointment in the treatment of veterans, Louis had great pride in his service to his country. He collected books on the Fifth Infantry Division and kept newspaper clippings that featured World War II and his division. He displayed his medals proudly, knowing he fought for freedom in a war that really meant something.

Louis was awarded two Bronze Star medals for individual acts of valor. He received the EAME ribbon with five Bronze Service Stars. He earned a Silver Service Star for the five campaigns he took part in: Normandy, Northern France, Rhineland, Ardennes-Alsace, and Central Europe. He was awarded the Croix de Guerre as a member of the Fifth Infantry Division by the French Government. He was also awarded a Combat Infantry Badge, Victory Medal, Conspicuous Service Cross and a Good Conduct Medal. He was very proud of his service to his country.

In 1998, he was honored and taken to the premier of *Saving Private Ryan* in a limousine by his local newspaper, *Newsday*. He was featured in their July 23, 1998, article "Witnesses to Combat," which describe the reaction of World War II veterans to the movie. On June 3, 2001, he was honored at a medal ceremony in Kings Point, Long Island, and awarded the Jubilee Medal of Honor. His family came with him to the ceremony and it was a very proud moment for Louis.

In 2001, he corresponded with family members of fellow GI, James Stephens, who lost his life in the war. Louis remembered him and gave his family great peace of mind with the information he told them. He told them how James died near Dornot on September 8, 1944. Louis never forgot the friends that didn't return from war and considered the rest of his life a gift from God.

During the war, Louis promised God that if he survived he would help others less fortunate than himself. He was always helping friends and other families with odd jobs. He usually devoted one day a week to helping someone else, in addition to working long, hard days himself. Over the years he worked multiple jobs as an auto mechanic, airplane mechanic, and plumber. After his retirement, he kept busy helping others with all types of work. He never called a repairman and was always able to fix everything.

Louis had two sons, Anthony in 1947 and Kenneth in 1955. His sons dearly loved and respected their father. Louis loved both his sons and made sure they both got the education that he wasn't able to pursue. Education was very important to Louis and he always regretted not being able to go any further in school. His wife still has his elementary school diploma in their safe, which he saved his entire life.

Louis's oldest son Anthony inherited his father's work ethic and learned how to fix everything and anything. He grew up helping his father fix cars and while they worked his father would tell him about the war. When Anthony was older, he and his father rebuilt a 1926 Nash together from a pile of parts. Anthony also built a large addition onto his house without any help from contractors and of course Louis was there to help.

Louis was very supportive of his children and even though money was tight he bought Ken a drum set and always encouraged his music career. Ken and his wife Amy lived together in a multi-family home with his parents. They were very close. Ken fondly remembers his father and writes:

> My earliest impression of my father, as a young boy, was a man who was very quiet, extremely strong and very hard working. He was an auto mechanic for most of his life.
> Every night at 5:30 or 6:00 P.M. I'd hear his VW Beetle rattle up the narrow driveway of our Queens home.
> His clothes were dirty, and hands stained dark from grease even after several washings. He'd walk up the steps into the kitchen and lean over to give my mother a kiss. He'd take his seat at the Formica table and wait for my mother to bring him his dinner and a cup of hot coffee. He was mildly superstitious and always sat in the same seat at the table. He never broke the routine. Sometimes my mother, brother and I would wait for him to eat but when he came home late we'd eat without him.
> Dad was average height, about 5' 4" but was built like a body builder. He had huge bulging forearms with veins ripping through them like lightning bolts. His hands were thick and meaty like a catcher's mitt. His arms developed

from many years lifting heavy peanut bags in the factory as a child and then years of manual labor as a mechanic.

He was tough and powerful. I've seen him pick up a VW engine, place in on his knees and install it in a VW Beetle without the use of a car jack. On a family vacation, my uncles and I witnessed him rip a coconut in half with his bare hands. My uncles affectionately named him Balu, after a gorilla we saw on the trip. The name stuck with him for years.

You could tell Dad was tired after an 8 to 10 hour day of working in a minimally heated garage repairing cars. His mood was usually quiet, but happy. Once in a while, he'd vent about his boss who was a bit of an under-handed character and the conversation was usually about a bounced check or money he was shorted but for some reason he stayed employed there for many years.

Occasionally, at dinner he'd have a small glass of wine that his father made. My grandfather produced what we called "rocket fuel." I never saw Dad drink more than a glass and I never saw him drunk or even tipsy; he never smoked and rarely gambled more than an occasional Lotto ticket or on a horse race when his brother called with a "hot tip."

After dinner, he'd have his cup of coffee that was usually overboiled to the point of being burnt. He drank it with a drop of milk and no sugar. He loved coffee and drank pots of it throughout the day back then.

Dad was always there to help no matter who you were. He had one philosophy, "Be good and honest to me and you have a friend for life." Dad would help anyone and never expected anything in return. I used to tease him when he did side jobs for his friends. I would ask him how much money he lost doing the job. Most of the time, he was happy if he recouped his supplies and gas money. There were times though, when people took advantage of his good nature and trust.

My friends loved him. He was easy to talk to. When they came over, I would often find them sitting on the front steps or in the backyard talking to him about music or just about anything you could think of.

Everyone he knew depended upon him for something. Even though he never had a formal education you could rely on him for anything and he'd figure it out. I'm still amazed at the worldly knowledge he possessed.

He'd sit out in the yard on summer nights and he could tell you what constellations you were looking at in the sky or what planets or stars were visible depending on the season. I always wondered how he knew them.

He could fix almost anything. With seven brothers, each involved in a trade, they all knew how to do everything from plumbing to carpentry, masonry, electrical work, car mechanics, and fixing machines and tools. It was an education that would be difficult to obtain in a school.

Sometimes it would amaze me how he would resolve mechanical or design problems. As I was growing up, he never called a repairman. If it was broken, he'd fix it, design it or build it from scratch.

Dad was even tempered. I rarely saw him angry. My parents rarely fought and when they did, it was over in minutes. No grudges held. They said what they had to say and then it was over as quickly as it started. That was his way. My mom could have stayed angry but Dad wouldn't let her. He'd make her laugh and then it was over.

Afterword

Louis was a devoted family man. He was a loving husband who made sure to tell his wife Olga he loved her at least once a day. They were happily married for 62 years. Louis is survived by five grandchildren, Kristin Gurley, Stephanie Lauria, Michelle Lauria, Kevin Lauria and myself, Amanda Anderson, and one great grandchild, Avery Anderson.

On November 8, 2008, Louis J. Lauria passed away. He had been in the hospital for over two months with complications from heart and kidney failure. His wife, son Ken, daughter-in-law Amy, brother-in-law Mike and sister-in-law Elsie visted him every day he was in the hospital and his son Anthony and his wife Linda drove up every weekend from Pennsylvania. Louis's family was by his side when he died. Ken describes his last moments:

> We continued to talk to him and noticed that he could hear us. We asked him questions and after a while we got a head nod or a raised eyebrow. At around 4:30 pm he opened his eyes. He was surrounded by his family. He looked at us. He seemed at peace. At 4:45 pm, his eyes slowly closed and his head turned to the side. I was looking at his breathing before and I could see that he was not breathing now. I called the nurse in the hallway and she came in and took his pulse. She examined him with a stethoscope and called the cardiac doctor on duty.
>
> The doctor told us what we already knew. Dad was gone and I can't think of a better way than to be surrounded by a family that loved him.
>
> I loved my dad. Like most people that knew him, I depended on him for everything. There is nothing in this home that doesn't remind me of him, from his tools to his garden, the rooms he built and everything that he fixed over the years. There is truly a void. But within that void, I am reminded every day of his presence. He will always be a hero—as a soldier and as a father.

We all miss him terribly.

Appendix
The Fifth Infantry Division in World War II

It is only fitting to include information about the Fifth Infantry Division along with my grandfather's memoir. The Fifth Infantry Division was extremely important to him and a great source of pride. Over the years, he collected articles, radio transcripts, pamphlets, and books about the Fifth Infantry Division and the 11th Infantry Regiment. He was also a member of the Society of the Fifth Division, United States Army. His notebooks and scrap papers often had his sketches of a red diamond on them, the division's emblem. When he helped make a concrete patio for his brother-in-law, Mike, without him knowing, he drew a giant diamond in the concrete right in the middle of the patio.

This section includes information about the Fifth Infantry Division, with special attention to the actions of the 11th Infantry Regiment. It also includes available information about radio and wire communications. Individual acts of valor of the Fifth Infantry Division are also highlighted, although they represent just a small sample of the heroic acts of the men of the Fifth Infantry Division. The section provides a general history of the Fifth Infantry Division in the ETO with special attention to events and battles that my grandfather mentions in his memoir. Since my grandfather's memoir is mostly of first-person experiences, this should help fill in historical information about the Fifth Infantry Division in World War II.

My grandfather was very proud of the accomplishments of the Fifth Infantry Division for good reason. They completed over 26 river crossings and covered over 2,049 miles in the European Theater Campaign.[1] The Fifth liberated 700 miles of France in just 27 days.[2] The Fifth Infantry Division's motto fittingly was, "We Will." General George S. Patton, commander of the Third Army, wrote the following about the Fifth Infantry Division:

Nothing I can say can add to the glory which you have achieved. Throughout the whole advance across France you spearheaded the attack of your Corps. You crossed so many rivers that I am persuaded many of you have web feet and I know that all of you have dauntless spirit. To my mind history does not record incidents of greater valor than your crossing of the Sauer and Rhine.[3]

Patton also remarked that the Fifth was a unit which, when told to "get up and get," "got up and got."[4] Major General S. Leroy Irwin, who commanded the Fifth Infantry Division for the majority of its combat tour in the ETO, remarked, "It is not only a great division, it is a magnificent one."[5] The red diamond Fifth Infantry Division played a pivotal role in liberating France and defeating Germany in Word War II.

Members of the Fifth Infantry Division wore a patch of a red diamond on their left sleeve. It stands for "intense loyalty, versatility of performance and the fulfillment of the motto 'We Will' when given a task to perform."[6] It originated in World War I, when Major Charles A. Meals of the Quartermaster Corps suggested that the emblem of the Fifth Division should be the "Ace of Diamonds, less the ace."[7]

The Fifth Infantry Division was inactive for 18 years after World War I. On October 16, 1939, units began to assemble at Fort McClellan, Alabama, under the command of Brigadier General Campbell B. Hodges.[8] The Fifth was considered a "triangular" division because it consisted of three infantry regiments: the Second, Tenth and Eleventh Infantries. The Fifth also contained four field artillery battalions: the 19th, 21st, 46th and 50th. Divisional troops consisted of the Seventh Combat Engineers, Fifth Quartermaster Battalion, Fifth Medical Battalion, Fifth Signal Company, Division Headquarters and Military Police Company.[9]

The division went through intensive training at Fort McClellan until the spring of 1940, when they reassembled at Camp McCoy in Wisconsin to train and study methods of combat in extreme cold weather conditions.[10] On September 4, 1940, Major General Joseph Cummins assumed command of the division and its headquarters were moved to Fort Custer, Michigan.[11]

On May 20, 1941, the division made a 600-mile march to central Tennessee, where it participated in Second Army maneuvers until June 28.[12] Next the division, with the exception of the Tenth Combat Team, went to Louisiana to participate in war game maneuvers in late August. The Tenth was on its way to Iceland, whose strategic importance made it a potential target for enemy attacks, and so the Fifth had orders to protect the island. The U.S. feared that German control of Iceland would allow them to use the island to refuel and make continental attacks on the U.S. With war looming in Europe, the Fifth prepared to go overseas. Once the rest of the division returned to Fort Custer, they embarked for Iceland, where they arrived from September 16, 1941, until May 10, 1942.[13] Most of the 11th Infantry Regiment sailed on April 7, 1941, aboard the *Orizaba* and the *Minargo*.[14]

With their arrival in Iceland beginning in 1941, the Fifth was the first division overseas in World War II in the ETO.[15] For the next 15 months, the Fifth endured severe climate conditions and long hours of hard work in Iceland.[16] The combat troops worked as deckhands, sodded Niessen huts, and helped build Meeks airfield on the Keflavik peninsula.[17] The Fifth trained in winter warfare techniques and did its best to acclimate to the limited daylight hours. GIs recalled that while in Iceland, "the entire division was garrisoning the perimeter of the island against German threats. Numerous German planes came over on reconnaissance, but the constant wind, work, and boredom were the hardest to fight."[18] The Fifth nicknamed themselves the "FBI," The Forgotten Bastards of Iceland.[19] With little regret, the Fifth finally left Iceland behind in early August 1943 and went on to Tidworth Barracks located in southcentral England.

A cannon company of 105-millimeter howitzers was added to each infantry regiment in October of 1943, which is when Louis Lauria joined up with the 11th Infantry Regiment. In England, training concentrated on familiarizing troops with weapons and making them experts in as many types as possible. They practiced with new equipment on the available range facilities.[20] On July 3, 1943, Major General S. Leroy Irwin assumed command and led the division until the final weeks of the war.[21] In late October 1943, the division moved to Northern Ireland and continued its extensive training at Camp Ballykindler and Camp Donard Lodge.[22]

In Ireland, the Fifth endured exhaustive training for eight months to thoroughly prepare itself for the imminent battle ahead. The division practiced river crossings on small Irish lakes where engineers trained to build multiple types of bridges.[23] The training proved invaluable for the 26 river crossings the division completed during combat.

The specialized units continued their training as well. The Fifth Signal Company was responsible for radio and telephone communication for the Fifth Division. Each cannon company also had its own radio wire crew in order to keep constant communication between the guns and the observation and command posts. Field radios and "walkie talkies" were new innovations of the time. Initially, field radios had a range of about five miles but required open areas to work successfully.[24] Buildings, overpasses, and bridges presented obstacles for the new technology.[25]

The troops continued to depend on telephone wires laid manually or by trucks to ensure communication. One Signal Corps officer explained the importance of telephone wires in combat:

> With radio you get the sets, issue them and hope for the best. They're wonderful of course we couldn't do without them but when you've got a good wire circuit in you're not kidding. It gives you direct instant and private communication. The enemy normally can't pick it up with intercept equipment or triangulate on the sending source. Atmospheric interference can't blot your messages out. Yes, wire may be hard to maintain, but it's awfully easy to talk over.[26]

The Signal Company and radio wire crew had a daunting task ahead of them. In their 10 months of combat, they used 18,780 miles of telephone field wire.[27] They had to repair wires under heavy combat conditions and completed mission objectives while carrying heavy D8–300 field radios and wire reels. Sometimes trucks hauled the field radios, but the radios often had to go where vehicles couldn't reach and so the crew had to carry them. Some field radios weighed as much as a hundred pounds.[28]

Throughout its training, the division also weathered seemingly constant rain. Men prepared for battle night and day on courses with overhead small-arms fire, minefields, and bursting demolition charges.[29] The artillery and anti–aircraft machine gunners trained to perfect their skills and the unit covered all details required for battle, including scouting, patrolling, map and aerial photograph reading, first aid, sanitation, camouflage, use of mines, booby traps, grenades and rockets.[30] When General Eisenhower and General Patton reviewed the 11th Infantry Regiment, Patton remarked, "This is the fittest, roughest, readiest outfit that I've ever inspected."[31]

D-Day arrived on Tuesday, June 6, 1944, and plans were issued for the Fifth Division's own arrival in France. On July 2, 1944, officers informed troops that they were restricted to their camps and billets and no longer able to fraternize with civilians because they were under secret departure orders.[32] The division left camp and headed for Belfast, Ireland, where they embarked for France on July 6, 1944. They boarded twelve Liberty ships and five passenger ships in all.[33]

Normandy

The Fifth arrived on the shores of Normandy on July 9, 1944. It landed off Les Dunes de Varreville on Utah Beach on the east coast of the Cherbourg Peninsula near St. Mère Église. A mist was in the air when the troops climbed down the side of the Liberty ships in full field gear, holding their duffle bags. Some troops waded ashore through three-foot-deep water while others disembarked on a steel ramp about a hundred yards out from shore.[34] Troops waded ashore onto Utah Sugar Red beach, which had been won 33 days before. The Germans were only 12 miles away and the beaches were still subject to frequent German air attacks.[35]

The men piled their duffle bags on the shore for later transport. The vehicles weren't unloaded until the next day and so the troops had to make a five-mile march in their wet boots to transit area B. After a short rest, they continued nearly 20 miles to reach the division concentration area near Montebourg. Upon landing, the Fifth Infantry Division was attached to the First Army under Lt. General Omar Bradley. Later, on the eleventh of July, they were assigned to the V Corps under Major General Leonard Gerow.[36]

The French countryside was a constant reminder of war, with devastated towns, dead livestock, and artillery exploding in the distance. There were signs everywhere stating "mines cleared to hedges" or "mines cleared to edge of road."[37] Normandy was hedgerow country. The landscape was covered by thick mazes of hedgerows which were designed to contain livestock in the fields.

The hedgerows had existed for hundreds of years. Thick mounds of dirt and rock were overgrown with trees, vines and brush, making nearly impassable barriers. For centuries, the hedgerows had presented difficulties for armies. Julius Caesar even remarked on the hindrance caused by hedgerows, stating, "These hedges present a fortification like a wall, through which it was not only impossible to enter, but even to penetrate with the eye ... the march of our army would be obstructed by these things."[38]

The hedgerows made visibility extremely difficult causing tactical dilemmas for the First Army. Infantry assault positions proved impossible because they couldn't see the enemy past the hedgerows.[39] Tanks were unable to drive through the thick undergrowth. Eventually, tactics were developed to tackle the hedgerows. While artillery suppressed the enemy machine-gun and anti-tank fire, engineers used explosives to blow holes in the hedgerow for the tanks and infantry to advance through.[40] Engineers also constructed plows and attached them to the front of tanks to burrow through the hedgerows.

The Fifth Division got its first taste of battle in Caumont, where for ten days they battled alongside the British 11th Armored Division and the Second Infantry Division against the German Third and Fifth Parachute Divisions.[41] The Fifth Infantry Division successfully held Caumont before it was relieved by the British on July 23, 1944. The Third Battalion of the 11th Infantry Regiment remained behind with the Second Infantry Division to occupy Hill 192, which was paramount for taking Saint-Lô.[42] The attack on Hill 192 was the result of over a week of planning, scouting and rehearsal of the combined forces of three divisions, the artillery and air support.[43]

The Fifth Infantry Division moved on to Cerisy la Forêt, where over three thousand American bomber planes and their escorts roared overhead toward Saint-Lô on July 25, 1944. The bombs prepared the way for the Saint-Lô breakthrough. During the heavy bombing, General McNair, chief of the army ground forces, was killed by bombs that fell short of the bomb line while he was observing the operation.[44] Lauria alluded to his death in his memoir. General Bradley requested the bombers fly in from the east to reduce friendly fire losses but most of the bombers of the Eighth Air Force came in from the north, killing 111 of their own men and wounding 490.[45]

Saint-Lô was nearly 90 percent destroyed by the air attack and later was referred to as "the Capitol of Ruins" by Samuel Beckett, the Nobel Prize–winning author of *Waiting for Godot*. The air attack was tremendously successful in paving the way for the breakthrough of the Third Army across France.

The bombing of Saint-Lô was pivotal in "Operation Cobra," where the Allies were able to create a gap in the German lines and break out of Normandy. The Fifth moved southeast and launched its first attack at Vidouville on July 26, 1944. It successfully secured Vidouville and Haut Vidouville.[46] On August 3, 1944, the Fifth Infantry Division was assigned to the XX Corps of the newly formed Third Army under Lt. General George S. Patton, Jr.

After the decimation of Saint-Lô, the Fifth began its drive across France as part of "Operation Cobra." Its first objective was to take the town of Angers. A radio broadcast on February 21, 1945, recounts the Fifth's heroic race across France:

Most people at home have the idea that Patton's sensational race across France was done by the armored division alone, but that isn't quite correct. The 5th Infantry Division was the spearhead of the attack a good deal of the time, and is very proud of the fact that it was never preceded by armor at any time except for a few minutes on August 31st, when three tanks got into Verdun an inch or two ahead of the first Red Diamond infantryman. Now how this was done simply defeats the imagination. I wasn't there and I find it hard to understand how a whole infantry division, with nothing but its own organic and assigned means of transport, could keep up a pace of 50, 60 and even 90 miles a day. The fact is that they did. They dropped everything to do it, left their kitchens and all their belongings behind; lived on K rations for 25 days; piled on to the tanks, tank destroyers, jeeps, trucks, ambulances or anything else they could find that would move. They were eight to ten on the outside of a medium tank, 12 to 14 clinging on to a tank destroyer.[47]

The Fifth dumped all unessential supplies in order to use its supply trucks to transport troops. The men were covered in dust from the journey and adapted to the constant travel. Some even managed to sleep as they held on to the vehicles.[48] The Fifth headed east from Saint-Lô to Coutances, turned south and then passed through Avranches.[49] The Air Corps' P-47s and P-51s preceded them, knocking out German resistance. The Germans scrambled to reorganize while the Air Corps left an aftermath of burned-out tanks, cars, trucks and dead livestock along the roads.[50] On August 7, 1944, the division received orders to take the city of Angers, with a population over 80,000.[51] The Fifth was to seize the bridges in Angers across the Maine and Loire rivers.

In the battle of Angers, fierce fighting developed over an intact railway bridge across the Maine River. While waiting for commands, an L Company patrol sneaked up and captured a German sentry who waited to detonate the bridge at any sign of attack.[52] The Germans embedded a massive amount of mines and explosives in the bridge. A railroad boxcar full of explosives stood in the middle of the bridge.[53] Even though the bridge was under fire, the Third Battalion crossed on the night of August 8, 1944.[54] The German infantry counterattacked and desperately tried to blow up the bridge. The Germans ran downhill towards the bridge, firing their machine guns and rifles in the dark. The Fifth fired back at the Germans, who had tied explosives around their waists to blow up the bridge but instead caused only their own fiery death.[55]

In Angers, the Fifth was able to hold the south bridge on the Maine River which split the city in two. The Fifth Infantry Division secured Angers in three days of fierce street fighting. During the battle of Angers, Lauria earned one of his Bronze Star Medals for his bravery in maintaining telephone communications to implement key attacks. The army record of the battle notes, "Communications were excellent, consisting of telephone wire lines to each battalion and divisions, supplemented by radio which worked effectively and was amplified by liaison officers."[56]

Angers was the first city of significant size that the Allies liberated.[57] French crowds greeted the troops with flowers, food and wine. Civilians tore down Nazi signs hanging throughout the city, including a large German officer's club banner.[58] The French civilians also burned posters of Hitler in the streets.[59] The troops discovered German wine stocks in Angers and found enough champagne to issue one bottle to every soldier. On August 11, 1944, Major General S. Leroy Irwin, Fifth Division commander, decorated 20 officers of the 11th Infantry, one of the 19th F.A. and two of the 725th Tank Battalion before a crowd of cheering French civilians.[60] Angers was a significant victory because Germans fleeing the Brest Peninsula had used Angers as their principal point of exit.[61]

On the twelfth of August, most of the Fifth Infantry Division left Angers and continued its drive across France through St. Calais and on to Chartres over 150 miles away. On their journey eastward, members of the division encountered jubilant French citizens grateful for the Allied troops who liberated their homes. One account explained, "Most of the way led through towns and villages where the French were going wild with joy, and the dirty, exhausted infantrymen were pelted with flowers."[62] They arrived in Chartres August 15 and 16.[63] The Seventh Armored Division had been fighting with its artillery for two days but encountered stiff resistance and was forced out of Chartres. The Army Corps changed tactics and thought the infantry was better suited to take the town, so the Fifth tackled the job of locating and flushing out the German infantry. Strategically, Chartres was pivotal in liberating France and was known as the "gateway to Paris."

The 11th Combat Team was given the mission of taking Chartres. The 11th Combat Team consisted of the 11th Infantry Regiment, Company C Fifth Medical Battalion, Company C of the Seventh Engineer Battalion, Company C 735th Tank Battalion, Company C 818th Tank Destroyer Battalion, Battery A of the 449th AAA-AW Battalion, and the 19th Field Artillery Battalion.[64] The Second and Third Battalions of the 11th Infantry Regiment located a pocket of German resistance. After a brief barrage from the Cannon Company, the 11th captured one hundred prisoners.[65] As the Third Battalion approached the city, it found another group of about one thousand Germans.

Captain H. M. Smith, Sgt. Clarence White and radio operator Paul Tredanari of the O.P. from Cannon Company were traveling with forward elements of F/2–11th Infantry and located this second pocket of resistance.[66] The men set up their observation post in a nearby barn. When the morning mist cleared, the men peered through a hole in the roof and spotted a 40-millimeter gun, a 20-millimeter gun and an 88-millimeter gun on a hundred yard front. The enemy was set up on the edge of woods facing a clearing with another strip of woods and a few buildings about 50 yards away.[67] The guns were in front of a church that had a large Red Cross flag draped across the front of it. [68]

Smith gave fire adjustments to Tredanari, who called back to command post on the SCR-300.[69] Lauria was back at the command post and radioed the fire mission to the guns. The 10-millimeter howitzers of Cannon Company hit the enemy observation post in the church steeple and then blasted the guns in the surrounding area with 200 rounds of high-explosive quick-fuse shells for around 45 minutes of constant fire.[70] The Germans fled into the surrounding woods into nearby foxholes. The shells that exploded in the trees had the "same effect as time fire scattering shrapnel downwards into the holes."[71] During the attack, they hit an ammunition dump and created a spectacular explosion like fireworks.

Once the Cannon Company's shelling let up, the infantry men were able to walk into the area with just four shots fired by Germans. The Germans surrendered the entire force of over 700 men. Between 200 and 300 Germans were killed in the attack. Four men of F Company lost their lives.[72] F Company also freed 47 men from A company who were captured earlier by Germans on August 17.[73]

Luckily, the Chartres cathedral was untouched by the battle. The Fifth liberated Chartres by August 19, 1944, and captured much-needed materials in German warehouses, including food, clothing and essential vehicles.[74] The 11th Infantry also confiscated a huge German limousine which they presented to Lt. General George S. Patton, who expressed his gratitude:

> You were very generous indeed to send me the lovely automobile captured by the Third Battalion, 11th Infantry Regiment, under Major Birdsong.... It will take a car as fast as that to catch up with you, which I hope to do shortly. I am deeply appreciative and full of admiration for the magnificent work you and the Division have done.[75]

The Fifth Infantry Division continued east toward Etampes, 34 miles from Chartres. Etampes was on the railroad line running south of

Paris and it supplied German troops in the south of France. It was well defended by German infantry units with machine guns, anti-aircraft and anti-tank 88-millimeter guns, roadblocks and minefields.[76]

On the night of August 21, three companies of the Fifth Infantry Division surrounded the city, cutting off escape routes to the north, southeast and east. Patrols entered the city and found unoccupied enemy positions.[77] The regiment with the Second Battalion and Third Battalion entered the city on the morning of the 22nd to find no resistance. The town was taken and Company A of the Seventh Engineers cleared the town of mines and booby traps.[78] After Etampes was liberated, French civilians confronted French women who had collaborated with the Nazis during their occupation. They cut their hair down to their scalps to shame them for their betrayal.[79]

The Fifth Infantry Division continued eastward, taking Malesherbes and the bridge over the Essones. It proceeded to take La Chapelle-la-Reine on August 23, where they captured 278 men and killed at least 75.[80] They also successfully secured an intact bridge at Nemours.[81]

Seine

On August 23, the 11th Combat Team, Fifth Cavalry Reconnaissance Troop, Regimental Intelligence, and Reconnaissance Platoon attacked Fontainbleau. The Germans blew the bridge across the Seine just as the 11th Combat Team arrived and dug in on the west side of the river.[82] It was essential that the Fifth Infantry Division establish a bridge across the Seine for Allied troops to continue across southern France.

The commander of the Second Battalion of the 11th Infantry, Lt. Col. Kelly B. Lemmon, spotted five small boats on the eastern side of the Seine and swam across to obtain them. He returned to the west side of the river under light enemy fire with all five boats. Captain Jack S. Gerrie of Company G and Platoon Sgt. Dupe A. Willingham secured a ten-foot canoe. They paddled across to the eastern shore, where Gerrie killed one German and was pinned down by retaliating fire and he ordered Willingham back to bring the men over.[83] Gerrie was able to swim back underwater to the west shore under heavy fire and called for a medium tank to fire on the location of the Germans on the far shore. The tank made an accurate attack which enabled the company to use the canoe and civilian boats to cross the river and establish a bridgehead.[84] Lemmon,

Gerrie and Willingham were awarded the Distinguished Service Cross for their heroic swims.[85]

With great effort, the 11th Combat Team was able to hold the bridgehead under heavy mortar and artillery fire. After intense fighting, the Company C engineers and the 1103 Engineer Group constructed a treadway bridge by 1630 hours on August 24 under heavy Germany artillery fire.[86] As soon as the bridge was completed, tanks, tank-destroyers and the Cannon Company howitzers rushed across, followed by the remaining companies of the First and Third Battalions of the 11th Infantry. The enemy withdrew on August 25 and the Fifth captured over 300 Germans and killed around 300.[87]

The Second Battalion of the 10th Infantry pushed onwards towards Montereau on the morning of August 25, 1944.[88] The Seine curves around to the east and so once again presented an obstacle for the troops. The 46th and 21st Field Artillery Battalions, 735th Tank Battalion, and 818th Tank Destroyer Battalion fired on the east bank of the Seine all day to prepare for another crossing while engineers constructed 70 assault boats to cross the river. The crossing was made under light artillery shelling at 2105 hours on August 25.[89]

A thick fog covered the area the next morning and when the men awoke, they surprised and captured 28 sleeping Germans with two 70-millimeter howitzers just on the other side of a garden wall. When the rest of the German force realized the Americans had crossed the Seine, they staged a counterattack through the dense fog that hid their position. Unfortunately for the Germans, the fog lifted suddenly, exposing their position in an open field. Around 60 Germans were killed once they became easy targets.[90] Approximately 100 Germans were forced out of the woods by artillery fire that exploded in the treetops. In the attack, 362 Germans were captured and several hundred were killed.[91] Twenty Americans lost their lives and 86 were seriously wounded.[92]

During this river crossing of the Seine, Medic Pvt. Harold A. Garman of Company B of the Fifth Medical Battalion earned the Congressional Medal of Honor for rescuing three injured men from the water. He was the only member of the Fifth Infantry Division to be awarded the Medal of Honor. During the crossing, injured men were evacuated to the southern shore in assault boats by their litter bearers from the medical battalion. Garman was on the friendly shore carrying the wounded from the boats to wait for ambulances.

He watched as a boatload of the wounded reached midstream and was assaulted by enemy machine-gun fire from about one hundred yards away. All of the men who were in the boat jumped in the water except one who was so badly injured that he was unable to get up from his litter. Two of the wounded were unable to swim and they clung to the side of the boat. Garman swam out to the men through ongoing machine-gun fire and towed the boat to the friendly shore. He saved the lives of the men. Other men were so inspired by his courage that they used additional assault boats to join the fight on the other side to save more wounded men.[93]

The Fifth pressed on and the Second Combat Team crossed the Yonne river and liberated Nogent, where they established yet another bridgehead northward across the Seine and constructed a treadway bridge. The Second RCT crossed the Marne River and liberated Reims between August 29 and 30.

The French civilians were jubilant at the arrival of American troops in Reims. They hopped onto tanks and trucks to thank the incoming troops. During celebrations in the street, one eager GI kissed every girl he could get his hands on.[94] Civilians made a large bonfire in the street and burned copies of *Mein Kampf* and SS pamphlets.[95] In Reims, civilians also punished French women who collaborated with the SS and cut their hair completely off.[96] The French citizens were relieved when the Americans arrived and they reported incidents of German abuses. They relayed that in Chatenay the Germans lined up and killed 18 French civilians for no apparent reason and in a nearby village they disemboweled the prettiest girl.[97]

Moselle

The Fifth continued moving eastward and the 11th Infantry Regiment captured Verdun on September 1, 1944. By now, the Fifth Infantry Division had traveled 700 miles in 26 days and had tragically run out of supplies. They were out of gas. The Fifth Infantry Division was dependent on gasoline for almost all essential action, including cooking and running ambulances, jeeps, tanks, cub planes and engineer equipment.[98] The Germans were still reeling from the onslaught of the Third Army's attack and had deserted Metz on September 3, 1944.[99]

As a result of the gas shortage, the Fifth Infantry Division was immobilized from September 1 until September 6 giving the Germans time to reorganize and return to Metz. The Germans were able to strengthen their forces on the east side of the Moselle. The Fifth Infantry Division experienced the most resistance and heaviest losses of the war in the next 25 days.[100]

The 11th RCT attempted to cross the Moselle at Dornot on September 7, 1944. They experienced 36 enemy counterattacks and had to withdraw from the beachhead after enduring over 60 hours of constant bombardment. The wounded were given strict orders not to cry out because they would disclose their position.[101] There were many casualties but the men obeyed the command and were silent when they were hit.

Bad weather prevented air support from aiding the 11th RCT. Due to heavy losses, an evacuation was ordered the night of September 10 at 2100 hours.[102] Men crossed back over the Moselle to the friendly shore by any means possible. Some men used empty water cans or empty ammunition cases to help them float across the river. Some pulled themselves along ropes attached to boats. Men helped each other make it across the river but many drowned in the cold water that night. Lt. Wright swam back to help swimmers make it ashore. He returned with some men and went out to help more but never was seen again.[103]

During the attack, Pfc. Dickey and Pfc. Frank Lalopa of K Company manned an outpost on a fringe of woods the first night when the Germans attacked. Their squad leader ordered them to withdraw but they refused. They fired their M-1s at the advancing Germans until they were killed. Twenty-two dead Germans were found in front of their position the next morning, just a few yards from their bodies.[104]

Some men were left behind on the east bank. They were either asleep or passed out when the withdrawal was ordered at 2100 hours. One man, Private Joseph Lewakowski, awoke the next morning to find the riverbank littered with dead Germans. He crawled over their bodies to the river and luckily found ropes enabling him to pull himself back to the friendly shore and reunite with his men.[105]

While the 11th drew the attention of the German artillery at Dornot, the 10th Regiment crossed the Moselle several miles south near Arnaville, about seven miles south of Metz.[106] The 10th Regiment was able to cross the night of September 9 without attracting the Germans. They advanced three hundred yards before the Germans noticed their position.[107] The Germans struck back with constant machine-gun, mortar, artillery and tank fire on the infantry assault units, the village of Arnaville and the bridgehead.[108]

The men of the 10th Infantry Regiment and the First and Third Battalions of the 11th Infantry Regiment did not have the support of anti-tank guns that were on the far bank of the Moselle and so the men counterattacked against the tanks using their rifles, machine guns, mortars, rifle grenades and hand grenades.

Meanwhile, engineers struggled to finish a bridge under constant artillery fire and worked under a smoke screen in order to provide some cover. A bridge was completed by noon on September 12, enabling supporting units to cross, including the Seventh Armored Division. The bridgehead was secure by September 15, 1944, at a great cost. Fourteen hundred men of the Fifth Infantry Division were either killed or wounded in the attempt to secure a bridgehead on the Moselle.[109] The 10th Infantry Regiment alone lost 24 officers and 674 men, which was 15 percent of their officers and 22 percent of their enlisted men.[110] Their bridgehead on the Moselle was the first permanent bridge secured for the Third Army and it would be essential for the successful attack on Metz in November.[111]

Metz

No attacking enemy had ever been able to capture Metz because it was extremely well defended and surrounded by fortresses. It was rich with history and had been transformed into a fortified city during the Roman Empire.[112] Twenty-two great forts and multiple bunkers constructed of steel-reinforced thick concrete encircled Metz.[113] The forts and bunkers had their own underground tunnels, power plants, water tanks and workshops, a strategic nightmare for any attacking army.[114] After the gas shortage, the Germans had time to return to Metz and prepare to hold the city. Fort Driant, the largest of the forts, was considered a key element in breaking through the defense to take Metz.[115]

On September 27, the first preliminary attack on Fort Driant was made by the 11th Infantry Division. Its attack, even with some light air support, failed to make an impact on the fort and so they withdrew that evening. They were counting on additional air support but bad weather at the take-off field prevented more planes from aiding them although the skies had cleared over Fort Driant.[116] Fort Driant had been inundated with cold rain for two consecutive weeks.[117] The Fifth regrouped to prepare for another attack. It focused efforts on making new demolition equipment to tackle the forts and bunkers, including satchel charges, pole charges, Bangalore torpedoes and "snakes," which were charges that were placed in long poles ahead of a tank.[118]

The next attack on Fort Driant began on October 3, 1944, by a tank company of the 735th, the Second Battalion, 11th Infantry and B, K and 11th Company of the Second Infantry Regiment.[119] They planned a preliminary attack of intense artillery barrage and napalm bombs. The tank-infantry team, supported by engineers, artillery and air fire, attacked the southwest and northwest corners of the fort using "snakes," flame throwers, Bangalore torpedoes, and pole and satchel charges. Air support used 500- and 1,000-pound high-explosive napalm bombs with delayed action fuses.[120] The attack was plagued by complications from disabled tank dozers, malfunctioning "snakes," poor road conditions, and sickness.

The men fought fearlessly above ground and underground against overwhelming odds. They were able to seize the southwestern third of the fort, including two concrete barracks, two subterranean connecting tunnels and perimeter pillboxes.[121] On October 12, 1944, the Third Army halted the attack because it couldn't take Fort Driant without sustaining devastating casualties.[122] The 95th Division relieved the Fifth Infantry Division on October 20 to defend the Moselle bridgehead, giving the Fifth its first chance to rest since landing in Normandy.

The Fifth moved near Luxembourg for a ten-day training period to prepare for attacking the forts. The Fifth endured almost 44 days of near-constant battle with the enemy on the Moselle before being moved to Piennes to rest, take showers, watch movies and most importantly train for the ensuing combat. They returned to the Moselle bridgehead on November 1 and relieved the 95th Division.

On November 8, 1944, the Third Army began its assault on the city of Metz when 1,300 bombers flew overhead, bombing towns, forts, and crossroads.[123] The Second Combat Team led the push through floodwaters. The Moselle and Seille Rivers were at their highest levels in over 30 years.[124] The Fifth was able to take Metz by avoiding the forts and concentrating its forces on liberating the city itself.[125] The Fifth coordinated its regiments for the attack. The Second Infantry drove east, the 10th Infantry pushed northeast and then north and the 11th was on the left flank and waited until the Second Infantry turned north before it joined the attack.[126] The Third Army experienced some of the strongest opposition and longest battles in Metz. They managed to isolate most of the forts by November 17 and concentrated on the city.

On November 18, 1944, two jeep loads of the 11th Infantry Regiment's communications crew made it into the heart of Metz, which was still enemy territory, by mistake.[127] They were trying to find the advanced command post to string their telephone wire but missed a turn and drove right into the main part of the city, which was held by Germans.[128] The Germans realized the men were Americans and began to fire. Luckily the men made it back alive.[129] There was an article in the *New York Times* about this event. There is a good chance Lauria was part of this group. He didn't mention it in his memoir but perhaps he never realized how close he had come to the heart of the German defenses.

On November 21, Metz officially surrendered, but the last fort didn't fall until December 8, 1944.[130] Not only did the troops have to face the heavily enforced forts at Metz, they battled mud, rain, flooding, snow, daily haze and rampant trench foot.[131]

Although Metz had surrendered, the battle continued because several forts held out. The men of the Fifth Infantry Division continued to fight with everything they had and there were countless acts of heroism. On November 29, 1944, 19-year-old medic Pvt. Duane N. Kinman of the Second Infantry Regiment saved the life of a rifleman who was wounded in the throat. Kinman could see he wasn't able to breathe and his trachea and throat muscles had been torn. Without anesthetic or surgical instruments, he preformed a tracheotomy with his pocket knife under intense shell fire in the blood-soaked, muddy battlefield. Before he made his first incision, he told his patient, "I don't like to do this, but it's the only way you are going to live."[132] After he made his incision in the trachea, Kinman inserted the top of the patient's fountain pen to keep his airway open.

The patient lived and was able to walk over to the aid station where he received a proper tracheotomy tube. At the field hospital, doctors inspected the tracheotomy site in amazement and didn't need to make any additional improvements. Word spread about Kinman's impressive surgery and the Western Reserve University of Ohio offered him a surgical scholarship after the war.[133]

After the battle of Metz, the 11th Infantry Division moved eastward about 30 miles from Metz and entered St. Avold on December 2, 1944, less than ten miles from the German border. There they found military barracks that had recently held German troops. Civilians warned an officer that the retreating Germans had planted at least 30 time bombs in the St. Avold area that were set to detonate after four days.

The military barracks were already partially occupied by an anti–

aircraft unit. The battalion officers searched the barracks for time bombs but were unable to locate anything. The next night, the First Battalion, Service Company and Cannon Company stayed in the military barracks. At 2300 hours on December 3, 1944, a time bomb exploded in the largest of the buildings, crushing and killing an unknown number of men beneath the collapsed building.[134] Engineers worked feverishly and managed to rescue some survivors.

The Fifth's four-month campaign in Metz and surrounding areas opened the door to the Sauer River and the Siegfried Line.[135] They were finally fighting Germans on their own soil. On December 5, 1944, the Fifth captured Lauterbach.[136] The Fifth's next objective was to take on the Saarlautern area which contained portions of the Siegfried Line.[137] It was diverted from further attacks on the Siegfried Line because the war demanded its attention elsewhere.

Battle of the Bulge

On December 16, 1944, Hitler began his initial counteroffensive in the Ardennes Forest. Patton called on the Fifth Infantry Division to relieve the exhausted Fourth Infantry Division, which was stationed one hundred miles to the northwest on the southern face of the Bulge to protect Luxembourg City.[138] American troops were decimated by the sudden onslaught of the massive German offensive. They were so desperate for reinforcements that clerks, cooks, drivers, and supply men found themselves on the front lines.[139] On December 20, 1944, the Fifth relieved the Fourth Division after traveling one hundred miles through snow and harsh winter conditions, completing the journey within 24 hours of receiving its orders.[140] The Fifth successfully protected Luxembourg City and pushed the Germans back, recapturing American equipment and supplies.[141] They also captured 830 German prisoners.[142]

Haller and Waldbillig returned to American hands by Christmas Day. The Fifth reached the southern bank of the Sauer River by New Year's Day, 1945.[143] The Fifth made a surprise crossing of the Sauer near Diekirch, Luxembourg, on January 18, 1945. Thanks to thorough reconnaissance, it was able to cross the river undiscovered and reach positions on high ground. By dawn, the entire company had crossed five hundred yards of open ground into German territory and completely surprised the enemy.[144]

After winning the bridgehead across the Sauer River in Luxembourg, the Fifth moved into Germany and led the drive of the XII Corps. It reached the Sauer River a second time near Echternach and Bollendorf. The 11th Combat Team crossed near Echternach and the 10th Combat Team crossed near Bollendorf.[145] The 10th and 11th Infantry Regiments both made their first attempt to cross the Sauer River between February 6 and 7, 1945, and encountered deadly resistance. Of the 38 assault boats used to reach the northern shore, all but two were destroyed. No one survived the attempted crossing.[146]

Swift currents contributed to the high casualties. Several boats capsized because of the rough water, barbwire and other underwater debris. The infantry men burdened with heavy equipment found swimming to safety impossible and drowned.[147] Newly constructed footbridges were also destroyed by strong currents from recent floods.[148] The river was swollen and twice its normal width. The Fifth continued to push to cross the Sauer and on February 10 the men were across the river. By the eleventh, engineers were able to construct a bridge that allowed tanks and artillery to cross.[149]

This portion of the Sauer River bordered Germany and was part of the Siegfried Line, which was a former World War I line of defensive bunkers and tank defenses bordering Germany. Hitler had reinforced the line prior to World War II, adding thousands of bunkers and pillboxes. The Siegfried Line stretched the entire length of the western German border and contained tens of thousands of bunkers, pillboxes and trenches.[150] The Fifth pushed through the Siegfried Line for the next ten days and crossed the Enz, Prum and Nims Rivers.[151]

A radio broadcast in February 1945 described the Fifth Infantry Division's heroic crossing of the Sauer:

> The high point of this recent action was the crossing of the Sauer river right smack into the Siegfried line. You have heard all about that—the high cliffs on each side of the river; the swift current of the stream; the German mortar and direct machine gun fire. How the G.I.s did it I will never be able to understand, but they did. And having made the crossing, they proceeded to crawl up those murderous hillsides, among the pillboxes, and capture them one by one. Sometimes they did it with dynamite and sometimes machine guns and rifles were enough ... don't allow anybody to tell you that the Siegfried line was easy or that it wasn't a remarkable thing to sail right into it that way, across a flooded river and into a honeycomb of cleverly concealed forts. Probably the Germans never dreamed that we would go into it right there. They no doubt expected us further down, where there aren't so many hills. But they've lost a big chunk of their line now, and every German infantry man experiences a

shock when he realizes it. You know the Germans have been convinced for some years now that nobody could ever take the Siegfried positions. The 5th Division, along with its friends and neighbors to left and right, have dealt with that superstition in the one way that counts.[152]

By February 26, 1945, the 11th Infantry was poised to attack Bitburg, Germany. They encountered little resistance and captured the town with some minor house-to-house fighting. By 0900 on February 28, 1945, most of Bitburg was secure. The Germans had lost the strong defenses of the Siegfried Line and scrambled to defend the Kyll River in disorganized groups. The Germans blew up bridges across the Kyll River in their hasty retreat. Steep, densely wooded hills bordered the Kyll River providing effective defense barriers and excellent observation points for the Germans.[153] The Kyll River assault by the Fifth Infantry Division was the easternmost Third Army position in Germany and it opened up the bridgehead for the Fourth Armored Division to dash towards the Rhine.[154]

The Fifth Infantry Division pushed onward towards the Rhine and crossed the Moselle River for the second time on March 13, 1945.[155] Assault, alligator and motor boats were utilized to complete the crossing while engineers constructed a pontoon bridge near Muden. The Fifth Infantry Division moved onwards and completed yet another river crossing.

On March 15, 1945, the 11th Combat Team approached Lieg, Germany. Here small-arms fire and shell fire as well as some tank-infantry attacks took place. This is where Lauria lost his friends and witnessed their horrific deaths. The 11th Infantry Regiment army records describe the incident as follows:

> Cannon Company crossed into the bridgehead in the morning and moved into firing position north of Lieg. While setting up a Fire Control Point, 1st Lt. Henry Bass, CO, and 1st Lt. Anderson, Executive Officer (who had transferred to Cannon Company from E Company on that very day), and Pfc. Owen R. Stanley, Company aid man, were all killed by direct artillery hit. Enemy shelling was particularly intense that morning. In addition to artillery batteries, the enemy was believed to have brought up a number of SP's and tanks.[156]

Rhine

The Fifth Infantry Division arrived at the Rhine River near Oppenheim on March 22, 1945. Farther north in the Remagen area the Ludendorf Bridge, a railway bridge, had been seized intact by the First Army but the area was under heavy attack by the Luftwaffe and German panzer divisions.[157] The day the Fifth Infantry Division arrived, General George S. Patton, Jr., gave it orders to cross at Oppenheim. The division had little time to scout the area for ideal locations for bridgeheads, but their lack of hesitation in the assault caught the Germans off guard. German east bank defenders were concentrated in an area farther south, near Worms, where they expected the Fifth to cross.

The 11th Infantry Regiment crossed the Rhine with two rifle battalions on the night of March 22, 1945.[158] It was the first assault crossing of the Rhine in history. Third Battalion's K Company arrived first without a single shot being fired, surprising seven Germans who quickly surrendered.[159]

By this time, the navy had set up headquarters at Metz and was able to send equipment necessary for river crossing forward. The Fifth Infantry Division received amphibious tanks, weasels, pontoon bridges, power launches, tank lighters and an endless amount of necessary equipment for amphibious operations.[160]

The 11th Infantry Regiment encountered some resistance from the Luftwaffe but was able to repel them with anti–aircraft guns. Hilly terrain provided an excellent view of enemy activity on the far shore, about 800 yards away.[161] The First Battalion hit the enemy with mortar fire to protect the crossing troops. Third Battalion's I Company was the second to cross and incurred increasingly heavy small-arms fire. Luckily, there were no casualties, but some of the men were wounded.

Seven hundred yards further down the river, First Battalion's A and C Companies met very heavy resistance establishing their bridgehead. They paddled directly into enemy fire while crossing a narrower section of the river 800 feet to the far shore. The men used rifle fire and grenades to punch a hole in the enemy's defense line of machine-gun nests. They used marching fire to expand the bridgehead and were able to strike inland.

Private First Class Paul Conn, Jr., stayed behind to hold down an enemy machine-gun nest. He dug in with his bare hands and spent the night alone less than 30 yards away from the machine-gun nest that shot at anything that moved. At sunrise, he spotted the nest of about 10 Germans and attempted to use his M-1, which unluckily jammed. In desperation, he pulled out a hand grenade and stood up before the enemy. The Germans were so surprised to see an American coming towards them that they threw up their arms and surrendered.[162]

The First Battalion received the brunt of the enemy resistance but the Fifth managed to surprise the enemy and the German artillery didn't respond until two hours after the initial crossing. The navy supplied landing craft and established a ferry service. The engineers constructed several pontoon and treadway bridges.

At 1815 hours on March 23, German planes few out of the haze and strafed the treadway bridge. Anti-aircraft gunners shot down three planes and no casualties or damages were incurred by the Americans. Forty-five minutes later German planes continued to make passes over the bridge every 15 minutes or so. One managed to hit a gasoline truck. The German planes continued strafing the bridge all night. The flares were dropped to conceal the troop position on the bridge.[163] Within 24 hours of the original crossing, the entire division had crossed and within 36 hours the bridgehead was secured. The bridgehead was five miles deep and seven miles wide.[164]

Main River

The Fifth Infantry Division continued northeast toward Frankfurt. The Fifth covered about 20 miles after crossing the Rhine and captured approximately 4,000 German prisoners along the way. Frankfurt was Germany's ninth-largest city and before the war had a population around 500,000. The British and Americans had bombed the city heavily and when the Fifth Infantry Division arrived only 100,000 civilians remained.[165] The primary obstacle in Frankfurt was crossing the Main River into the city over a rare intact bridge. Of all the bridges the Fifth encountered, it was the most heavily defended.[166]

At 1935 hours on the night of March 26, 1945, the Third Battalion of the 11th Infantry Regiment raced across the bridge at the Main River under heavy artillery fire. The bridge was about five hundred yards long and under constant bombardment from Germans, who battered the bridge with everything they had. The command post of the Third Battalion was hit nine times by heavy shelling. Despite the overwhelming odds, by 0700 hours the next day the battalion had established a bridgehead six hundred yards deep and eight hundred yards wide.[167] Frankfurt had a large railway station and was home to significant airfields that the Germans fought desperately to hold.

In Frankfurt, the Fifth encountered fierce house-to-house fighting. Civilians with binoculars directed heavy mortar fire, making matters worse.¹⁶⁸ The Germans pinpointed where the troops were bringing in supplies from the bridgehead and hit the area with extremely heavy artillery fire which later the Fifth learned came from high-velocity anti-aircraft guns mounted on towers around Frankfurt. Casualties from this attack were heavy.

All battalions crossed the bridge into Frankfurt under heavy artillery fire. Throughout the city, troops encountered deadly sniper fire and difficult house-to-house fighting from the river to the railway station. Intact communications were paramount in coordinating a successful attack. The Fifth Infantry Division's historical record notes that "during this period wiremen worked frantically to repair lines frayed by the constant pounding of enemy shells."[169] It took four days to clear Frankfurt of German resistance.[170]

The city of Frankfurt was cleared by March 30. That day the Fifth Infantry Division was switched from the XII Corps to the XX Corps reserve of the Third Army. This gave the men their first chance for rest since November 1, 1944, near Metz. Although Frankfurt was completely destroyed, the men enjoyed a chance to rest from the constant fighting and seemingly endless movement eastward. The weather improved and men took advantage of the time to bathe, change into clean, dry clothes and clean their equipment. The men continued training exercises for the remainder of the war that loomed ahead.

Ruhr Pocket

On April 7, 1945, the division prepared to join the III Corps of the First Army further north in the Ruhr pocket. Troops traveled around one hundred miles north to the towns of Ostwig, Borghausen, Bigge and Nuttlar.[171] Several strong German divisions remained in the Ruhr pocket and the Fifth's objective was to check each village and quell any resistance. On April 10, the Fifth Infantry Division seized and crossed the Wenne River. It was the only unopposed river crossing they encountered in the entire war.[172] In the Ruhr pocket, an increasing number of Germans began to surrender. Multiple U.S. divisions surrounded the enemy, meeting erratic resistance. It wasn't uncommon for an infantry battalion to take around five hundred prisoners a day.[173] The Germans who continued fighting in the Ruhr pocket aimed to keep the Allies occupied and away from other action in Germany.

By this point in the war, some of the German commanders had lost their fanaticism due to their continuing losses. The German major who defended Visbeck was told to defend the town to the last man. He informed the SS that he wasn't up for the job but they instructed him to follow orders and die for Hitler. Before the war, he had been a transportation corps major in Paris with no military experience. He read in a military book how to defend the town and supervised the process of digging in. After two days and one sleepless night supervising the details, he went to bed. He was awakened at 0200 hours by a soldier of the F Company and forced to surrender. He was quite indignant that the 10th Infantry would attack at such an ungodly hour and protested because that type of warfare was not in his book.[174]

There were a few cases of Nazi soldiers who still fought with unflagging fanaticism. In the town of Schwammenberg, the 737th Tank Battalion located nine German officers holed up in a lodge. There they captured Lieutenant General Kurt von Kortzfleisch, who thrust his hand out in the Nazi salute and yelled out, "Heil Hitler!" three times. Then he proceeded to yell, "I am a true German fighting to the last.... I am glad to die for my Fuhrer." He dove for the woods and was quickly shot and killed. This is the only known account of a German officer who preferred death to surrendering to the Americans.[175]

On April 9, 1945, the Fifth Division uncovered a huge collection of "priceless Jewish manuscripts, paintings and other cultural artifacts" in Hungen and in the salt mines in Merkers.[176] The Germans had stolen from Jewish museums and collections from all over the world. The Fifth Division discovered gold bullion worth approximately $100,000,000 (in 1945) and priceless artwork including some by Degas.[177] Dr. Alfred Rosenberg, the German propagandist, used the collection in his biased research against Jewish culture. The army pledged to return the collection to its rightful owners, which included the Oppenheim Museum in Frankfurt and the Jewish Historical Museum in Amsterdam.[178]

By April 16, 1945, the Ruhr pocket was secured and cleared of enemy resistance, but the Fifth Infantry Division faced a new problem of processing the thousands of German prisoners of war. Most German officers gave their men orders not to surrender and to escape by any means possible. German soldiers tried to disguise themselves, usually unsuccessfully, as civilians. The troops set up roadblocks, searching for escaping soldiers. In the town of Arnsberg, all male civilians were rounded up and around 10 percent of them were German soldiers.[179]

The Fifth Infantry Division also learned that there were thousands of displaced slave laborers in the Ruhr pocket. Poles, Russians, Serbs, French, Greeks, Czechs, and Hungarians had been forced into slave labor. Thousands of recently freed prisoners took to the roads, not sure where to go or what to do. The Fifth Infantry Division helped set up camps and provided food and shelter for displaced persons.

On April 20, 1945, General Irwin was promoted as the new commander of the XII Corps. He had led the Fifth Infantry Division since June 3, 1943, and had been with them since training in Iceland. Major General Albert E. Brown took over command of the Fifth Infantry Division. He was an experienced general and a World War I veteran. When Irwin left the Fifth Infantry Division on April 22, 1945, he said:

> It is a blow to me to be separated from the unit whose fortunes I have shared since the last days in Iceland. We have seen much since then, and have made more than a little history. I cannot adequately express the pride and admiration I have for the Division. It is more than a great Division-it is a magnificent one. The skill and courage of its officers and men have carried the Division to such fame that its exploits are known throughout our country.[180]

On April 25, the Fifth Infantry Division rejoined the XII Corps of the Third Army and headed to Regen, near the Germany-Austria-Czechoslovakia border. Berlin fell to Russian hands by the end of April and Adolf Hitler committed suicide on April 30, 1945. The war was coming to a close but there were still pockets of resistance by Germans who continued to fight because they hadn't received orders to do otherwise.

By now, the Fifth Infantry Division traveled like experts. It moved over three hundred miles to Czechoslovakia despite a truck shortage.[181] It requested permission to use German vehicles that were abandoned all over the roads. The Fifth Infantry Division made the trip in two days and once again got aboard anything they could find to make the journey.

The Fifth Infantry Division set out to clear the area eastward in southern Czechoslovakia and northern Austria of enemy resistance. It encountered light resistance and was able to advance quickly despite mountainous terrain and cold, rainy weather conditions. Prisoners continued to eagerly surrender to the American troops, rushing westward to avoid being captured by the Russians.

In Volary, Czechoslovakia, the Fifth Infantry Division discovered a horrific act of unimaginable brutality. SS guards had led a group of Jewish men and women on a death march over seven hundred kilometers from concentration camps in eastern Poland. Along the way, they were constantly beaten and forced to live off grass and rotten potatoes.[182] The guards shot those men and women who were too exhausted to continue. The Allied advance prevented further movement and so the march ended in Volary. The men and women who died in Volary were buried in a shallow common grave. Over 80 bodies were exhumed by German civilians under the direction of the Fifth Infantry Division. Many had died of starvation while others had been murdered.

The medics of the Fifth Infantry Division took the 60 female survivors immediately to a German civilian hospital, where they were cared for by the medics. They were emaciated and suffered from severe malnutrition. In addition, many had tuberculosis, typhus, heart trouble, blistered and gangrenous feet, festered wounds, diarrhea, and frostbite.[183] The women were nursed back to health in the hospital in Volary.

The reconnaissance platoon of the 803rd Tank Destroyer Battalion moved out of Volary at 0820 hours on May 7, 1945. The German 11th Panzer Division ambushed the unit. Pfc. Charles Havlat of Dorchester, Nebraska, was shot through his helmet and died instantly.[184] He was the last GI of the Fifth Infantry Division to die in combat in the ETO. Ten minutes later, at 0831, a courier rushed up with the message "CEASE FIRE."

The war in the ETO had finally come to an end. The full message read, "Cease fire and cease all forward movement."[185] That evening in celebration, "men continued to fire guns, but this time not in anger. Ack-ack batteries shot red tracers into the starry sky, flares of all colors blazed trails across the blue, and on the ground bonfires dotted the landscape, silhouetting the Bohemian hills."[186]

It had been a long, arduous journey of over 2,000 miles for the Fifth Infantry Division from Normandy to Czechoslovakia. They accomplished many significant and heroic deeds but also suffered a tremendous loss of life. One hundred seventeen officers and 2,103 enlisted men in the Fifth Infantry Division were killed in action.[187] Eighteen officers and 319 enlisted men died of wounds and 2 enlisted men died of injuries.[188] The individual stories of each man may never be known, but they gave their lives selflessly to protect others, a true definition of a hero.

Men in the Fifth Infantry Division suffered thousands of wounds. Thirty-five officers and 698 enlisted men were seriously wounded in action and 355 officers and 7,158 enlisted men were slightly wounded in action.[189] One officer and 5 enlisted men were seriously injured in action and 44 officers and 1,091 enlisted men were slightly injured in action. All in all during the ten months of combat, 564 officers and 11,822 enlisted men were either killed or wounded in combat.[190] Four hundred fifty-three officers and 12,569 enlisted men suffered from non–battle conditions such as battle fatigue, trench foot and frostbite.[191]

Although the war was over, the Fifth Infantry Division was needed to help stabilize the postwar chaos. Refugees, displaced persons, and former prisoners of concentration camps and forced labor camps had no access to transportation, sanitation or food. Thousands of Germans citizens were homeless because their cities had been destroyed during combat. There was no remaining government to oversee order in the chaotic postwar civilization.

The Fifth Infantry Division helped carry out "Plan Eclipse" in southeastern Bavaria, helping establish postwar order. The first objective was to separate some 60,000 German troops who fled to American lines and return them to the Russians, their rightful captors.[192] The Fifth Infantry Division acclimated to a life without constant combat. For some, it was a difficult transition, as was coming to terms with everything they had experienced over the last year.

With the exception of a short break in late October and few days rest in early April, the Fifth Infantry Division endured 300 days of constant combat. Many soldiers started to lose their nerves once the combat ended and they had time to process everything they had encountered. The army recognized that in "prolonged engagements where the going is really tough" anywhere from 30 to 50 percent of soldiers requiring medical care suffered from combat exhaustion.[193]

German, French and American soldiers all suffered from some type of combat exhaustion.[194] The United States Army was aware of the medical problem and took action to prevent men from breaking down by making sure troops received adequate rest. For the Fifth Infantry Division, rest was a luxury. When soldiers were overwhelmed by the stresses of war, the medics were trained to use barbiturates to induce narcosis.[195] Sleep therapy between 24 and 72 hours in duration was the best medical intervention of the time to relieve soldiers' hysteria and eventually return them to duty.[196]

On June 18, 1945, the first units of the Fifth Division left Velshofen, Germany, and arrived in Camp Saint-Lôuis, an assembly area for homeward-bound troops on June 21, 1945.[197] From there, they took a three-day train trip to Le Havre Port of Embarkation to board ships bound for the U.S.

The 11th Infantry Division, 705th Infantry Regiment, 705th Ordinance Company, Fifth Medical Battalion, Fifth Reconnaissance Troop, Seventh Engineer Battalion and Fifth Signal Company boarded the crowded U.S.S. *Lejeune*, a former passenger ship.[198] In the *Diamond Drift*, a newsletter to troops, Captain F. W. MacDonald told the troops, "We regret that we have had to squeeze you into a ship which carried, when operated commercially, about one-fifth of the number of passengers. We know however, that you will be consoled by the knowledge that only one out of five would be getting home at this time if you traveled with all the comforts of home."[199] The men were thrilled to be going home but thought they would have to return to duty and fight in the Pacific.

Col. Paul J. Black addressed the men of the 11th Infantry Division on their journey home:

> To the Officers and Enlisted Men of the Regiment:
> It has been my privilege to have served with many of you through months of hazardous combat and wherever our missions were assigned in the field. Being that arm which has the final word in actually closing with and destroying the enemy, the Infantry has justified the faith placed in it. As members of the army, the men of this regiment, both old and new, in every rank have acquitted themselves proudly. By faithful and exemplary performance of duty you have contributed to the outstanding victories already won and have earned this chance to rejoin family and friends.
> I sincerely trust your leave will be as pleasant as you have anticipated and that when we assemble again our energies will be renewed to carry through to a swift and successful conclusion the tasks that remain before us. I am proud of the regiment and its achievements and feel sure that your conduct on furlough will not detract from its splendid record. I am confident that the spirit you have shown in combat will hasten the day when we can all return to our homes permanently with the knowledge of a job well done.[200]

The division landed in the U.S. at the end of July and was given a 30-day leave for "rest, recuperation, rehabilitation, and recovery." The war in the Pacific ended on September 2, 1945, and the division was inactivated on September 20, 1945.

The Fifth Infantry Division received comparatively little press during and even after the war for all its accomplishments. The men made key river crossings that were often the focal point of the Third Army's plans.[201] It was frequently the first to reach important rivers, including the Kyll and Rhine. Often press of the Fifth's actions didn't exist because it operated under a "security blackout" to increase its ability to take the enemy by surprise.[202]

Today there are only a few sources documenting the significant role that the Fifth Infantry Division played in the ETO. The Fifth Infantry Division had a pivotal role in the Allies' success in the ETO. Major General S. Leroy Irwin conveys due admiration for the Fifth Infantry Division: "to the brave and skilful men of all ranks and grades who made possible this story of success I wish to express my deep admiration, affection and respect. There are no finer soldiers in the world today."[203]

Chapter Notes

Introduction

1. Thomas Kessner, *The Golden Door: Italian and Jewish Immigrant Mobility in New York City, 1880–1915* (New York: Oxford University Press, 1977), 34.
2. John Lauria, interview by editor Amanda Page Anderson (and all subsequent interviews are by the editor), Howard Beach, NY, January 3, 2010.
3. Rose Lauria, interview, Queens, NY, December 29, 2000.
4. Louis Lauria, interview, East Northport, NY, December 28, 2000.
5. *Ibid.*
6. *Ibid.*.
7. John Lauria, interview, Howard Beach, NY, January 10, 2010.
8. Louis Lauria, interview, East Northport, NY, December 28, 2000.
9. Rose Lauria, interview, Queens, NY, December 29, 2000.
10. John Lauria, interview, Howard Beach, NY, January 3, 2010.
11. *Ibid.*
12. John Lauria, interview, Howard Beach, NY, December 29, 2000.
13. John Lauria, interview, Howard Beach, NY, January 3, 2010.
14. Louis Lauria, interview, East Northport, NY, December 28, 2000.
15. John Lauria, interview, Howard Beach, NY, January 3, 2010.
16. Kate Lauria, interview, Queens, NY, December 29, 2000.
17. *Ibid.*
18. *Ibid.*
19. John Lauria, interview, Howard Beach, NY, January 3, 2010.
20. Army Qualification Record, form no. 100, Louis J. Lauria, July 15, 1944, p. 1.
21. John Lauria, interview, Howard Beach, NY, January 3, 2010.

Appendix: The Fifth Infantry Division in World War II

1. Society of the Fifth Division, United States Army, "Fifth Infantry Division: World War II," http://www.societyofthefififthdivision.com/WWII/WW-II.htm.
2. U.S. Army Public Relations Offifice, *The Fifth Division in France* (Metz, France: L'Imprimerie Du Journal Le Lorrain, 1944), 3.
3. Society of the Fifth Division, United States Army, "Fifth Infantry Division: World War II," http://www.societyofthefififthdivision.com/WWII/WW-II.htm.
4. U.S. Army Public Relations Offifice, *The Fifth Division in France* (Metz, France: L'Imprimerie Du Journal Le Lorrain, 1944), 3.
5. Fifth Division Historical Section, *The Fifth Infantry Division in the ETO* (Atlanta: Albert Love Enterprises, 1945), 3.
6. *Ibid.*, 32.
7. *Ibid.*
8. *Ibid.*, 6.
9. *Ibid.*
10. *Ibid.*
11. Society of the Fifth Division, United States Army, "Fifth Infantry Division: World War II," http://www.societyofthefififthdivision.com/WWII/WW-II.htm.
12. Fifth Division Historical Section, *The Fifth Infantry Division in the ETO* (Atlanta: Albert Love Enterprises, 1945), 6.

13. *Ibid.*, 7.
14. U.S. Department of the Army, *Eleventh Infantry Regiment: Fifth Infantry Division* (Baton Rouge, LA: U.S. Department of the Army, 1946), 4.
15. Joseph W. Allen, Herman L. Bogart, and Vergil Miller, "The Saga of Seven Countries," *Diamond Drift*, July 19, 1945.
16. U.S. Department of the Army, *Eleventh Infantry Regiment: Fifth Infantry Division* (Baton Rouge, LA: U.S. Department of the Army, 1946), 4.
17. Fifth Division Historical Section, *The Fifth Infantry Division in the ETO* (Atlanta: Albert Love Enterprises, 1945), 7.
18. Joseph W. Allen, Herman L. Bogart, Vergil Miller, "The Saga of Seven Countries," *Diamond Drift*, July 19, 1945.
19. Edward J. Barta, *The Red Diamond's First Fifty: A History of the Fifth Infantry Division, 1917–1967* (Fort Carson, CO: Information Offifice, 1967), 17.
20. Fifth Division Historical Section, *The Fifth Infantry Division in the ETO* (Atlanta: Albert Love Enterprises, 1945), 7.
21. Society of the Fifth Division, United States Army, "Fifth Infantry Division: World War II," http://www.societyofthefififthdivision.com/WWII/WW-II.htm.
22. U.S. Department of the Army, *Eleventh Infantry Regiment: Fifth Infantry Division* (Baton Rouge: U.S. Department of the Army, 1946), 4.
23. Fifth Division Historical Section, *The Fifth Infantry Division in the ETO* (Atlanta: Albert Love Enterprises, 1945), 7.
24. U.S. Department of the Army, *Radio Sets SCR-509 and SCR-510* (Washington, DC: U.S. Department of the Army, 1943), 3.
25. *Ibid.*
26. John H. Walker, "Messengers of Battle," *Popular Science* 142 (June 1943): 52.
27. Fifth Division Historical Section, *The Fifth Infantry Division in the ETO* (Atlanta: Albert Love Enterprises, 1945), 244.
28. John H. Walker, "Messengers of Battle," *Popular Science* 142 (June 1943): 51.
29. Fifth Division Historical Section, *The Fifth Infantry Division in the ETO* (Atlanta: Albert Love Enterprises, 1945), 8.
30. *Ibid.*
31. U.S. Department of the Army, *Eleventh Infantry Regiment: Fifth Infantry Division* (Baton Rouge: U.S. Department of the Army, 1946), 4.
32. U.S. Army Public Relations Offifice, *The Fifth Division in France* (Metz, France: L'Imprimerie Du Journal Le Lorrain, 1944), 5.
33. Fifth Division Historical Section, *The Fifth Infantry Division in the ETO* (Atlanta: Albert Love Enterprises, 1945), 8.
34. U.S. Army Public Relations Offifice, *The Fifth Division in France* (Metz, France: L'Imprimerie Du Journal Le Lorrain, 1944), 6.
35. Fifth Division Historical Section, *The Fifth Infantry Division in the ETO* (Atlanta: Albert Love Enterprises, 1945), 41.
36. U.S. Army Public Relations Offifice, *The Fifth Division in France* (Metz, France: L'Imprimerie Du Journal Le Lorrain, 1944), 6.
37. *Ibid.*
38. The Internet Classics Archive, "*The Gallic Wars* by Julius Caesar," Daniel C. Stevenson, Web Atomics, http://classics.mit.edu/Caesar/gallic.2.2.html.
39. James Carafano, *After D-Day: Operation Cobra and the Normandy Breakout* (Boulder, CO: Lynne Rienner Publishers, 2000), 28.
40. Ibid, 42.
41. U.S. Army Public Relations Offifice, *The Fifth Division in France* (Metz, France: L'Imprimerie Du Journal Le Lorrain, 1944), 6.
42. U.S. Department of the Army, *Eleventh Infantry Regiment: Fifth Infantry Division* (Baton Rouge: U.S. Department of the Army, 1946), 6.
43. James Carafano, *After D-Day: Operation Cobra and the Normandy Breakout* (Boulder, CO: Lynne Rienner Publishers, 2000), 43.
44. U.S. Department of the Army, *Eleventh Infantry Regiment: Fifth Infantry Division* (Baton Rouge: U.S. Department of the Army, 1946), 6.
45. Andrew Williams, *D-Day to Berlin* (London: Hodder, 2005), 182.
46. Fifth Division Historical Section, *The Fifth Infantry Division in the ETO* (Atlanta: Albert Love Enterprises, 1945), 43.

47. Vincent Sheean, "Fifth Infantry Division," radio broadcast, February 21, 1945, 2.
48. *Ibid.*, 3.
49. U.S. Army Public Relations Offifice, *The Fifth Division in France* (Metz, France: L'Imprimerie Du Journal Le Lorrain, 1944), 9.
50. *Ibid.*
51. *Ibid.*, 11.
52. U.S. Department of the Army, *Eleventh Infantry Regiment: Fifth Infantry Division* (Baton Rouge: U.S. Department of the Army, 1946), 8.
53. *Ibid.*
54. U.S. Army Public Relations Offifice, *The Fifth Division in France* (Metz, France: L'Imprimerie Du Journal Le Lorrain, 1944), 12.
55. *Ibid.*
56. U.S. Department of the Army, *Eleventh Infantry Regiment: Fifth Infantry Division* (Baton Rouge: U.S. Department of the Army, 1946), 11.
57. U.S. Army Public Relations Offifice, *The Fifth Division in France* (Metz, France: L'Imprimerie Du Journal Le Lorrain, 1944), 12.
58. Tyler Alberts, *5th Infantry Division, Liberation of Western Europe*, series 2, volume 9, *Unedited Raw Combat Footage*, DVD (Fort Worth, TX: Combat Reels, 1944).
59. *Ibid.*
60. U.S. Department of the Army, *Eleventh Infantry Regiment: Fifth Infantry Division* (Baton Rouge: U.S. Department of the Army, 1946), 12.
61. Society of the Fifth Division, United States Army, "Fifth Infantry Division: World War II," http://www.societyofthefififthdivision.com/WWII/WW-II.htm.
62. Vincent Sheean, "Fifth Infantry Division," radio broadcast, February 21, 1945, 3.
63. U.S. Army Public Relations Offifice, *The Fifth Division in France* (Metz, France: L'Imprimerie Du Journal Le Lorrain, 1944), 13.
64. Fifth Division Historical Section, *The Fifth Infantry Division in the ETO* (Atlanta: Albert Love Enterprises, 1945), 71.
65. U.S. Department of the Army, *Eleventh Infantry Regiment: Fifth Infantry Division* (Baton Rouge: U.S. Department of the Army, 1946), 12.
66. Fifth Division Historical Section, *The Fifth Infantry Division in the ETO* (Atlanta: Albert Love Enterprises, 1945), 71.
67. *Ibid.*
68. *Ibid.*
69. U.S. Department of the Army, *Eleventh Infantry Regiment: Fifth Infantry Division* (Baton Rouge: U.S. Department of the Army, 1946), 12.
70. Fifth Division Historical Section, *The Fifth Infantry Division in the ETO* (Atlanta: Albert Love Enterprises, 1945), 71.
71. *Ibid.*
72. U.S. Army Public Relations Offifice, *The Fifth Division in France* (Metz, France: L'Imprimerie Du Journal Le Lorrain, 1944), 16.
73. U.S. Department of the Army, *Eleventh Infantry Regiment: Fifth Infantry Division* (Baton Rouge: U.S. Department of the Army, 1946), 13.
74. Fifth Division Historical Section, *The Fifth Infantry Division in the ETO* (Atlanta: Albert Love Enterprises, 1945), 71.
75. U.S. Department of the Army, *Eleventh Infantry Regiment: Fifth Infantry Division* (Baton Rouge: U.S. Department of the Army, 1946), 92.
76. Fifth Division Historical Section, *The Fifth Infantry Division in the ETO* (Atlanta: Albert Love Enterprises, 1945), 71.
77. U.S. Army Public Relations Offifice, *The Fifth Division in France* (Metz, France: L'Imprimerie Du Journal Le Lorrain, 1944), 6.
78. Fifth Division Historical Section, *The Fifth Infantry Division in the ETO* (Atlanta: Albert Love Enterprises, 1945), 71.
79. Tyler Alberts, *5th Infantry Division, Liberation of Western Europe*, series 2, volume 9, *Unedited Raw Combat Footage*, DVD (Fort Worth, TX: Combat Reels, 1944).
80. Fifth Division Historical Section, *The Fifth Infantry Division in the ETO* (Atlanta: Albert Love Enterprises, 1945), 75.
81. U.S. Army Public Relations Offifice, *The Fifth Division in France* (Metz, France:

L'Imprimerie Du Journal Le Lorrain, 1944), 17.
82. Fifth Division Historical Section, *The Fifth Infantry Division in the ETO* (Atlanta: Albert Love Enterprises, 1945), 71.
83. *Ibid.*, 77.
84. *Ibid.*, 78.
85. *Ibid.*
86. Society of the Fifth Division, United States Army, "Fifth Infantry Division World War II Individual Decorations," http://www.societyofthefififthdivision.com/WWII/ww2medals.htm.
87. U.S. Army Public Relations Offifice, *The Fifth Division in France* (Metz, France: L'Imprimerie Du Journal Le Lorrain, 1944), 18.
88. *Ibid.*
89. *Ibid.*
90. *Ibid.*
91. *Ibid.*
92. *Ibid.*
93. U.S. Department of the Army, *Eleventh Infantry Regiment: Fifth Infantry Division* (Baton Rouge: U.S. Department of the Army, 1946), 15.
94. Robert Hargis, Starr Sinton, and Ramiro Bujeiro, *World War II Medal of Honor Recipients (1), Navy & USMC* (Oxford: Osprey, 2003), 18.
95. Tyler Alberts, *5th Infantry Division, Liberation of Western Europe*, series 2, volume 9, *Unedited Raw Combat Footage*, DVD (Fort Worth, TX: Combat Reels, 1944).
96. *Ibid.*
97. *Ibid.*
98. U.S. Department of the Army, *Eleventh Infantry Regiment: Fifth Infantry Division* (Baton Rouge: U.S. Department of the Army, 1946), 15.
99. Fifth Division Historical Section, *The Fifth Infantry Division in the ETO* (Atlanta: Albert Love Enterprises, 1945), 92.
100. U.S. Army Public Relations Offifice, *The Fifth Division in France* (Metz, France: L'Imprimerie Du Journal Le Lorrain, 1944), 20.
101. Society of the Fifth Division, United States Army, "Fifth Infantry Division: World War II," http://www.societyofthefififthdivision.com/WWII/WW-II.htm.
102. U.S. Army Public Relations Offifice, *The Fifth Division in France* (Metz, France: L'Imprimerie Du Journal Le Lorrain, 1944), 22.
103. *Ibid.*
104. U.S. Department of the Army, *Eleventh Infantry Regiment: Fifth Infantry Division* (Baton Rouge: U.S. Department of the Army, 1946), 24.
105. U.S. Army Public Relations Offifice, *The Fifth Division in France* (Metz, France: L'Imprimerie Du Journal Le Lorrain, 1944), 23.
106. U.S. Department of the Army, *Eleventh Infantry Regiment: Fifth Infantry Division* (Baton Rouge: U.S. Department of the Army, 1946), 24.
107. Society of the Fifth Division, United States Army, "Fifth Infantry Division: World War II," http://www.societyofthefififthdivision.com/WWII/WW-II.htm.
108. U.S. Army Public Relations Offifice, *The Fifth Division in France* (Metz, France: L'Imprimerie Du Journal Le Lorrain, 1944), 24.
109. *Ibid.*
110. Society of the Fifth Division, United States Army, "Fifth Infantry Division: World War II," http://www.societyofthefififthdivision.com/WWII/WW-II.htm.
111. Fifth Division Historical Section, *The Fifth Infantry Division in the ETO* (Atlanta: Albert Love Enterprises, 1945), 112.
112. U.S. Army Public Relations Offifice, *The Fifth Division in France* (Metz, France: L'Imprimerie Du Journal Le Lorrain, 1944), 26.
113. Fifth Division Historical Section, *The Fifth Infantry Division in the ETO* (Atlanta: Albert Love Enterprises, 1945), 118.
114. U.S. Army Public Relations Offifice, *The Fifth Division in France* (Metz, France: L'Imprimerie Du Journal Le Lorrain, 1944), 28.
115. *Ibid.*, 27.
116. U.S. Department of the Army, *Eleventh Infantry Regiment: Fifth Infantry Division* (Baton Rouge: U.S. Department of the Army, 1946), 29.
117. *Ibid.*

118. "Entrance to Metz Fort Taken," *New York Times*, November 10, 1944.
119. Fifth Division Historical Section, *The Fifth Infantry Division in the ETO* (Atlanta: Albert Love Enterprises, 1945), 119.
120. U.S. Army Public Relations Offifice, *The Fifth Division in France* (Metz, France: L'Imprimerie Du Journal Le Lorrain, 1944), 27.
121. Fifth Division Historical Section, *The Fifth Infantry Division in the ETO* (Atlanta: Albert Love Enterprises, 1945), 120.
122. U.S. Army Public Relations Offifice, *The Fifth Division in France* (Metz, France: L'Imprimerie Du Journal Le Lorrain, 1944), 27.
123. *Ibid.*
124. *Ibid.*, 28.
125. *Ibid.*
126. *Ibid.*
127. Fifth Division Historical Section, *The Fifth Infantry Division in the ETO* (Atlanta: Albert Love Enterprises, 1945), 131.
128. "Two American Jeeps Reach Metz' Center," *New York Times*, November 20, 1944.
129. *Ibid.*
130. *Ibid.*
131. Fifth Division Historical Section, *The Fifth Infantry Division in the ETO* (Atlanta: Albert Love Enterprises, 1945), 134.
132. *Ibid.*, 133.
133. Gene Currivan, "Foxhole Surgeon, 19, Saves Soldier with Jackknife and Fountain Pen," *New York Times*, December 3, 1944.
134. U.S. Army Public Relations Offifice, *The Fifth Division in France* (Metz, France: L'Imprimerie Du Journal Le Lorrain, 1944), 29.
135. U.S. Department of the Army, *Eleventh Infantry Regiment: Fifth Infantry Division* (Baton Rouge: U.S. Department of the Army, 1946), 47.
136. Society of the Fifth Division, United States Army, "Fifth Infantry Division: World War II," http://www.societyofthefififthdivision.com/WWII/WW-II.htm.
137. Edward J. Barta, *The Red Diamond's First Fifty: A History of the Fifth Infantry Division, 1917–1967* (Fort Carson, CO: Information Offifice, 1967), 25.
138. *Ibid.*
139. *Ibid.*, 26.
140. Fifth Division Historical Section, *The Fifth Infantry Division in the ETO* (Atlanta: Albert Love Enterprises, 1945), 144.
141. *Ibid.*
142. Edward J. Barta, *The Red Diamond's First Fifty: A History of the Fifth Infantry Division, 1917–1967* (Fort Carson, CO: Information Offifice, 1967), 25.
143. Society of the Fifth Division, United States Army, "Fifth Infantry Division: World War II," http://www.societyofthefififthdivision.com/WWII/WW-II.htm.
144. Edward J. Barta, *The Red Diamond's First Fifty: A History of the Fifth Infantry Division, 1917–1967* (Fort Carson, CO: Information Offifice, 1967), 27.
145. Fifth Division Historical Section, *The Fifth Infantry Division in the ETO* (Atlanta: Albert Love Enterprises, 1945), 157.
146. *Ibid.*, 170.
147. Edward J. Barta, *The Red Diamond's First Fifty: A History of the Fifth Infantry Division, 1917–1967* (Fort Carson, CO: Information Offifice, 1967), 27.
148. Fifth Division Historical Section, *The Fifth Infantry Division in the ETO* (Atlanta: Albert Love Enterprises, 1945), 171.
149. Edward J. Barta, *The Red Diamond's First Fifty: A History of the Fifth Infantry Division, 1917–1967* (Fort Carson, CO: Information Offifice, 1967), 27.
150. *Ibid.*
151. Neil Short and Chris Taylor, *Germany's West Wall: The Siegfried Line* (Oxford: Osprey, 2004), 12.
152. Edward J. Barta, *The Red Diamond's First Fifty: A History of the Fifth Infantry Division, 1917–1967* (Fort Carson, CO: Information Offifice, 1967), 28.
153. Vincent Sheean, "Fifth Infantry Division," radio broadcast, February 21, 1945, 3–4.
154. U.S. Department of the Army, *Eleventh Infantry Regiment: Fifth Infantry Division* (Baton Rouge: U.S. Department of the Army, 1946), 67.
155. *Ibid.*
156. Fifth Division Historical Section, *The Fifth Infantry Division in the ETO* (Atlanta: Albert Love Enterprises, 1945), 193.

157. U.S. Department of the Army, *Eleventh Infantry Regiment: Fifth Infantry Division* (Baton Rouge: U.S. Department of the Army, 1946), 73.
158. Fifth Division Historical Section, *The Fifth Infantry Division in the ETO* (Atlanta: Albert Love Enterprises, 1945), 198.
159. *Ibid.*
160. *Ibid.*
161. *Ibid.*
162. *Ibid.*
163. *Ibid.*, 200.
164. *Ibid.*, 204.
165. Society of the Fifth Division, United States Army, "Fifth Infantry Division: World War II," http://www.societyofthefififthdivision.com/WWII/WWII.htm.
166. Fifth Division Historical Section, *The Fifth Infantry Division in the ETO* (Atlanta: Albert Love Enterprises, 1945), 208.
167. *Ibid.*
168. *Ibid.*
169. *Ibid.*
170. *Ibid.*, 209.
171. Edward J. Barta, *The Red Diamond's First Fifty: A History of the Fifth Infantry Division, 1917–1967* (Fort Carson, CO: Information Offifice, 1967), 29.
172. Fifth Division Historical Section, *The Fifth Infantry Division in the ETO* (Atlanta: Albert Love Enterprises, 1945), 212.
173. *Ibid.*, 225.
174. *Ibid.*,, 213.
175. *Ibid.*
176. *Ibid.*
177. "Jewish Art Cache Found in Germany: Third Army Seizes Collection Used for Propaganda by Reich Government," *New York Times*, April 10, 1945.
178. *Ibid.*
179. *Ibid.*
180. Fifth Division Historical Section, *The Fifth Infantry Division in the ETO* (Atlanta: Albert Love Enterprises, 1945), 213.
181. *Ibid.*
182. *Ibid.*, 218.
183. *Ibid.*, 220.
184. International Jewish Cemetery Project, "Volary: Prachatice, Bohemia," International Association of Jewish Genealogical Societies, http://www.iajgs.org/cemetery/czechrepublic/volary.html.
185. Society of the Fifth Division, United States Army, "Fifth Infantry Division: World War II," http://www.societyofthefififthdivision.com/WWII/WWII.htm.
186. Fifth Division Historical Section, *The Fifth Infantry Division in the ETO* (Atlanta: Albert Love Enterprises, 1945), 221.
187. "Third Army Bonfifires Mark GI's Reaction," *New York Times*, May 9, 1945.
188. Fifth Division Historical Section, *The Fifth Infantry Division in the ETO* (Atlanta: Albert Love Enterprises, 1945), 224.
189. *Ibid.*
190. *Ibid.*
191. *Ibid.*
192. *Ibid.*
193. *Ibid.*
194. United States Army Pictorial Service, *Combat Exhaustion*, DVD (Washington, DC: War Offifice, 1943).
195. *Ibid.*
196. *Ibid.*
197. *Ibid.*
198. Fifth Division Historical Section, *The Fifth Infantry Division in the ETO* (Atlanta: Albert Love Enterprises, 1945), 224.
199. F. W. MacDonald, "Letter to Passenger on U.S.S. LeJeune," *Diamond Drift*, July 19, 1945.
200. *Ibid.*
201. Paul J. Black, "Letter to Passenger on U.S.S. LeJeune," *Diamond Drift*, July 19, 1945.
202. Fifth Division Historical Section, *The Fifth Infantry Division in the ETO* (Atlanta: Albert Love Enterprises, 1945), 186.
203. *Ibid.*
204. U.S. Army Public Relations Offifice, *The Fifth Division in France* (Metz, France: L'Imprimerie Du Journal Le Lorrain, 1944), 1.

Works Cited

Alberts, Tyler. *5th Infantry Division, Liberation of Western Europe*, series 2, volume 9, *Unedited Raw Combat Footage*, DVD. Fort Worth, TX: Combat Reels, 1944.

Allen, Joseph W., Herman L. Bogart, and Vergil Miller. "The Saga of Seven Countries." *Diamond Drift*, July 19, 1945.

Army Qualifification Record. Form no. 100. Louis J. Lauria, July 15, 1944.

Barta, Edward J. *The Red Diamond's First Fifty: A History of the Fifth Infantry Division, 1917-1967*. Fort Carson, CO: Information Offifice, 1967.

Black, Paul J. "Letter to Passenger on U.S.S. LeJeune." *Diamond Drift*, July 19, 1945.

Carafano, James. *After D-Day: Operation Cobra and the Normandy Breakout*. Boulder, CO: Lynne Rienner Publishers, 2000.

Currivan, Gene. "Foxhole Surgeon, 19, Saves Soldier with Jackknife and Fountain Pen." *New York Times*, December 3, 1944.

"Entrance to Metz Fort Taken." *New York Times*, November 10, 1944.

Fifth Division Historical Section. *The Fifth Infantry Division in the ETO*. Atlanta: Albert Love Enterprises, 1945.

Hargis, Robert, Starr Sinton, and Ramiro Bujeiro. *World War II Medal of Honor Recipients (1), Navy & USMC*. Oxford: Osprey, 2003.

International Jewish Cemetery Project. "Volary: Prachatice, Bohemia." International Association of Jewish Genealogical Societies. http://www.iajgs.org/cemetery/czechrepublic/volary.html (accessed January 28, 2010).

Internet Classics Archive. "*The Gallic Wars* by Julius Caesar." Daniel C. Stevenson, Web Atomics, http://classics.mit.edu/Caesar/gallic.2.2.html (accessed January 15, 2010).

"Jewish Art Cache Found in Germany: Third Army Seizes Collection Used for Propaganda by Reich Government." *New York Times*, April 10, 1945.

Kessner, Thomas. *The Golden Door: Italian and Jewish Immigrant Mobility in New York City, 1880-1915*. New York: Oxford University Press, 1977.

MacDonald, F. W. "Letter to Passenger on U.S.S. LeJeune." *Diamond Drift*, July 19, 1945.

Sheean, Vincent. "Fifth Infantry Division." Radio broadcast, February 21, 1945.

Short, Neil, and Chris Taylor. *Germany's West Wall: The Siegfried Line*. Oxford: Osprey, 2004.

Society of the Fifth Division, United States Army. "Fifth Infantry Division: World War II." http://www.societyofthefififthdivision.com/WWII/WW-II.htm (accessed January 2, 2010).

"Third Army Bonfifires Mark GI's Reaction." *New York Times*, May 9, 1945.

"Two American Jeeps Reach Metz' Center." *New York Times*, November 20, 1944.
United States Army Pictorial Service. *Combat Exhaustion*. DVD. Washington, DC: War Offifice, 1943.
U.S. Army Public Relations Offifice. *The Fifth Division in France*. Metz, France: L'Imprimerie Du Journal Le Lorrain, 1944.
U.S. Department of the Army. *Eleventh Infantry Regiment: Fifth Infantry Division*. Baton Rouge, LA: U.S. Department of the Army, 1946.
U.S. Department of the Army. *Radio Sets SCR-509 and SCR-510*. Washington, DC: U.S. Department of the Army, 1943.
Walker, John H. "Messengers of Battle." *Popular Science* 142 (June 1943): 49–53.
Williams, Andrew. *D-Day to Berlin*. London: Hodder, 2005.

Index

American Almond Products 14, 15, 164, 173
Angers 49-61, 184-186
Ardennes, Luxembourg 104, 106
Arnaville 191

Bastogne 102
Battle of the Bulge 101-110, 195-196
Belfast, Ireland 28
Bitburg 125-129, 197
Bradley, Gen. Omar 102, 183, 184
Brown, Maj. Gen. Albert E. 202

Camp Ballykindler 181
Camp Donard Lodge 181
Camp McCoy 180
Camp Myles Standish 25
Camp Rucker 25
Camp Saint-Lôuis 165
Camp Upton 20-23
Caumont 60, 184
Chartres 62-69, 166, 186-187
Chatenay 190
Cherbourg 33
code 32, 36-37, 46, 64, 70
combat exhaustion 83, 136, 162-164
Cummins, Maj. Gen. Joseph 180

D-Day 30, 32
Diekirch 117, 195
Dornot 77-88, 89, 190-191

Echternach 196
80th Infantry Division 103
82nd Airborne Division 23
Eisenhower, Gen. Dwight 102
Etampes 187-188

Fifth Infantry Division 29, 155-156, 179-206
First Army 45, 183, 197, 200
Fontainebleau 70-74, 188-189

Fort Bragg 23-25
Fort Custer 180 181
Fort Driant 95, 192
Fort Fisher 33
Fort McClellan 180, 181
Fort Meade 25
Fourth Armored Division 103, 143, 155, 197
Fourth Infantry Division 104, 106
Frankfurt 149-154, 199
fuel shortage 94, 190

Garman, Pvt. Harold A. 189
Gerow, Maj. Gen. Leonard 183
Gerrie, Capt. Jack S. 188

Haller 111, 195
hedgerow 42-43, 49, 183
Hodges, Brig. Gen. Campbell 180
Hoscheid 120-122
Hungen 201

Irwin, Maj. Gen. Leroy 180, 186, 202, 206

Keflavik peninsula 29, 181
Kinman, Duane N. 194
Kyll River 197, 205

Lemmon, Kelly 188
Lieg 135, 141, 197
living conditions 68, 89, 106-108, 113, 120, 122
Luxembourg city 103, 111

Maginot Line 89
Main River 149-154, 199
mal'occhio 10
Marsico Nuovo 7, 8, 17
McNair, General 44, 184
medical treatment 40-41, 54, 82, 101, 194

215

Meekers 201
Messerschmitt 109, 144
Metz 87, 89–97, 190–195
Montereau 71–74, 189
Moselle River 89, 93–95, 140, 190, 192, 194
mouse hole fighting 53, 73, 151–152

95th Division 103

105 Howitzer 24, 29, 37, 38, 123–124, 181
Operation Cobra 184
Oppenheim 143–49, 197

P-47 70–72, 94, 132–133, 185
Patton, Gen. George 92, 102–103, 143, 179, 180, 182, 184, 187, 198
Peckler, Carl 14, 15, 164
Plane Eclipse 204
prisoners of war: Allies 131, 146–48; German 74, 201, 202

radio wire crew 34–39, 110, 182
ration (K & C) 31, 43, 52, 68, 185, 105, 153
Reagan, Ronald 127
Reims 74–75, 190

Remagen Bridge 149–154
Rhine River 135, 140, 143–145, 180, 197–99, 205
river crossings 49, 66, 96–97, 111–124, 198
Ruhr Pocket 154, 156–157

Saarlautern 101–104
Saint Avold 97–100, 194–5
Saint-Lô 43–45, 184, 195–197
Saint Mere 33
Sauer River 195–197, 111–124, 180, 195
Seine River 71–74, 188–190
Siegfried Line 195, 196
slave labor 202

telephone wire repair 52, 113–114, 182, 200
35th Division 24

Verdun 75–76, 190
Vidouville 184
Volary 202–203

Wallerstädten 145
Willingham, Sgt. Dupe A. 188
Wright, Lieutenant 191

www.ingramcontent.com/pod-product-compliance
Ingram Content Group UK Ltd.
Pitfield, Milton Keynes, MK11 3LW, UK
UKHW041955140426
5217IPUK00015B/807